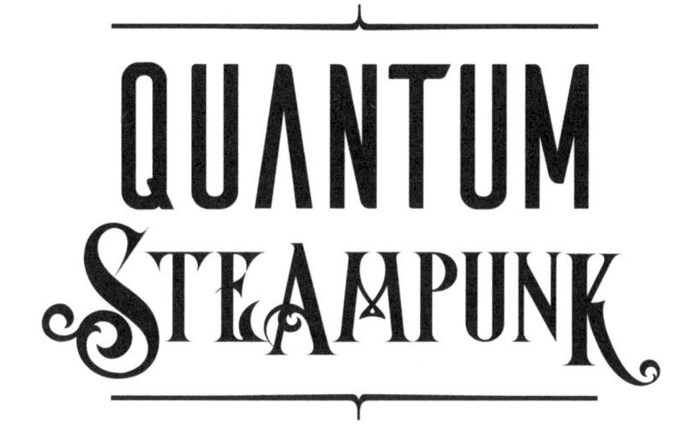
QUANTUM
STEAMPUNK

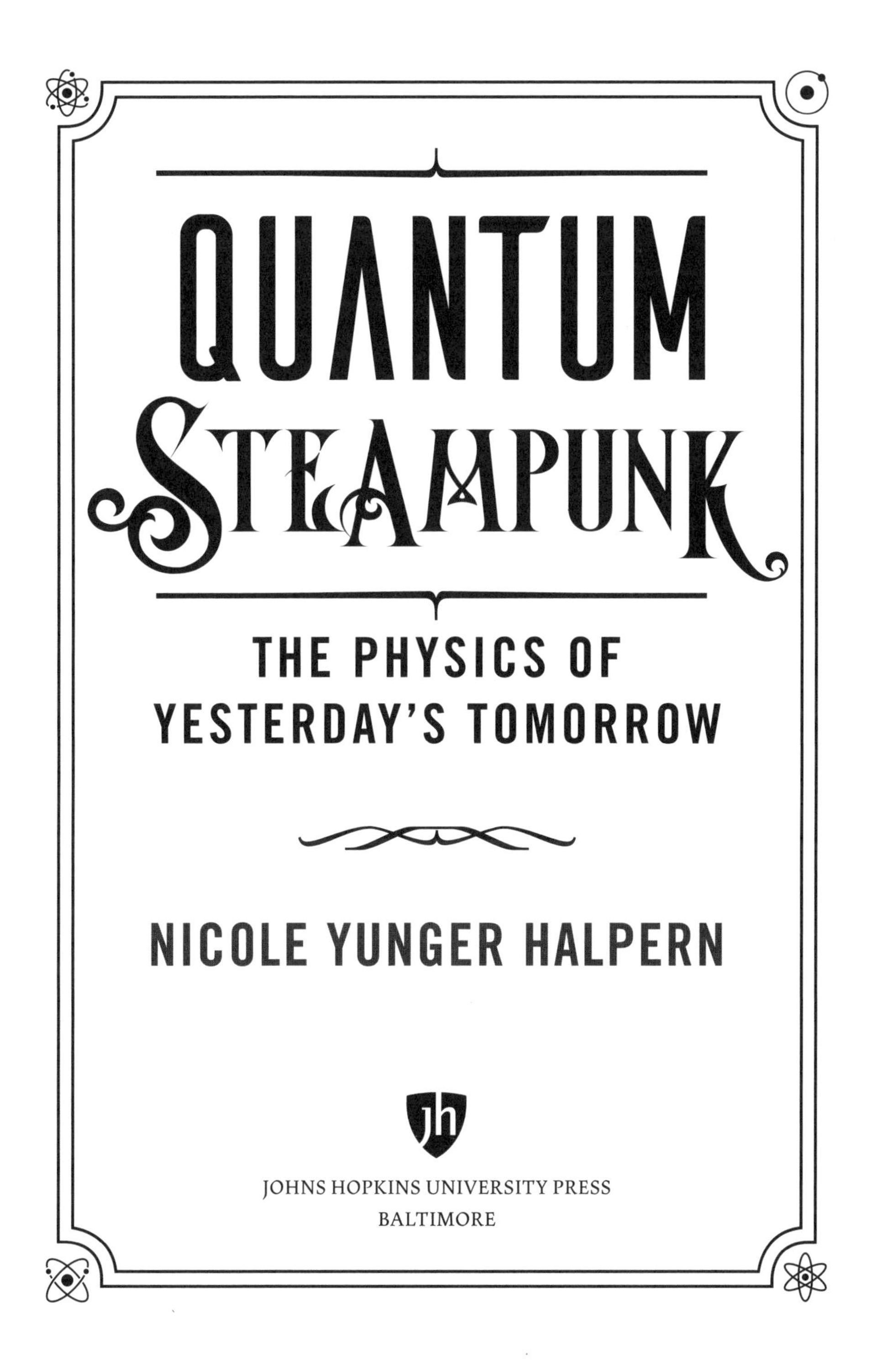

QUANTUM STEAMPUNK

THE PHYSICS OF YESTERDAY'S TOMORROW

NICOLE YUNGER HALPERN

JOHNS HOPKINS UNIVERSITY PRESS
BALTIMORE

Printed in the United States of America on acid-free paper
2 4 6 8 9 7 5 3 1

•

Johns Hopkins University Press
2715 North Charles Street
Baltimore, Maryland 21218-4363
www.press.jhu.edu

•

Library of Congress Cataloging-in-Publication Data

Names: Yunger Halpern, Nicole, 1989– author.
Title: Quantum steampunk : the physics of yesterday's tomorrow / Nicole Yunger Halpern.
Description: Baltimore : Johns Hopkins University Press, 2022. | Includes bibliographical references and index.
Identifiers: LCCN 2021028550 | ISBN 9781421443720 (hardcover) | ISBN 9781421443737 (ebook)
Subjects: LCSH: Quantum theory. | Thermodynamics.
Classification: LCC QC174.12 .Y86 2022 | DDC 530.12—dc23
LC record available at https://lccn.loc.gov/2021028550

•

A catalog record for this book is available from the British Library.

•

All illustrations by Todd Cahill.

•

Special discounts are available for bulk purchases of this book.
For more information, please contact Special Sales
at specialsales@jh.edu.

Thanks to
my parents · and · brother
for helping to ensure
that I'd
become a reader
who could
become a writer

CONTENTS

QUANTUM
STEAMPUNK

CHAPTER 0

PROLOGUE

ONCE UPON A TIME IN PHYSICS

"Forgive me for arriving late, Audrey." A shadow detached itself from the doorway's other shadows and stepped into the drawing room. The firelight played over a man in his mid-twenties, looking like a raven who's tumbled into a ditch. One wouldn't believe that the cloak he'd handed the butler had borne the brunt of the storm. "Our friend Ewart—ah, arranged for me to be delayed."

"'Our friend,' indeed." A girl, about ten years younger than Caspian, shut her book and beckoned him into the warmth. "You have never flattered a soul in your life, Caspian, so pray do not begin with Ewart. Daisy insisted on bringing out the tea things before you arrived," she added as Caspian squelched his way across the red-and-gold carpet. "Would you care for a cup?"

"Please." Caspian stretched his hands out toward the fire as a raindrop trickled into his right ear. "I trust that your brother is well?"

"If any unwellness were ill-informed enough to approach Baxter, he would turn a cartwheel, and it would sail straight past." Audrey set down the teapot. "Sugar?"

"If you would be so kind." Caspian turned from the fire to settle himself into the high-backed chair beside the table, as only a long-time family friend or a cat can in another's house. "You said that you had a discovery to report." He leaned closer. "A discovery about time, and about rewriting time on the quantum scale."

Looking up to pass Caspian his cup and saucer, Audrey saw a spot flicker by his right cheek—and not because of the firelight. Frowning, she set the china down.

"Indeed," she said, feeling around the table without taking her eyes off the spot. Her hand met a miniature crystal vase, which she emptied of its violets and water. "Precisely why I invited you." The flicker hovered closer to Caspian's cheek. "In fact—"

Thwack! Audrey clapped the vase's mouth to Caspian's right cheek and clapped a hand to his left cheek, immobilizing his head. Caspian flinched, and his eyes grew wide, but he didn't move otherwise.

"Keep still," Audrey hissed, "and hold this." Caspian took hold of the vase while she fetched a sheet of paper from a nearby table. She slid it under the vase, then lifted the vase away from his face, pressing the paper against the vase's mouth. The two of them stared at the vase in the firelight as a small, copper body buzzed against the glass.

"That is no insect," Caspian murmured.

Audrey shook her head.

"Whilst flattered by Ewart's eagerness to overhear my discovery," she said, "I could do without the compliment." Audrey set the vase upside-down atop the table and peered more closely. "I could never have caught this spy-fly of his if it had not been moving so slowly, but it must be nearly drained of energy."

"Poor thing," Caspian murmured. Audrey smiled.

"Baxter is developing a superior version, which can extract energy from a certain type of light quite efficiently. Not the type here," she added, waving at the fireplace, "but a type that Ewart would have in his laboratory."

The two of them watched the mechanical buzzing insect awhile longer, the fireplace crackling behind them, till a raindrop trickled down Caspian's nose.

"I neglect my duties as hostess," Audrey said, reaching for his cup and saucer. "Tea?"

AUDREY AND CASPIAN belong to a steampunk novel that lives in my imagination. Steampunk is a genre that stretches from literature to art and film. H. G. Wells and Jules Verne sowed the movement's seeds in nineteenth-century novels such as *The Time Machine* and *Twenty Thousand Leagues under the Sea*. In steampunk works, technologies of the future populate nineteenth-century settings. You'll find Victorian parlors, factories belching smoke, Western American saloons, and a Meiji Japan reinventing itself. Men wearing top hats and women wearing bustles engineer dirigibles, clockwork robots, and time machines. Steam—sheepdog of the Industrial Revolution—drives the fantasies' technologies. Yesterday and tomorrow blend into an aura of nostalgia, adventure, and romance.

Steampunk fans write novels, paint and draw, work metal and glass, and film movies. They don cravats and corsets, array themselves with telescopes and magnifying glasses, and gather at conventions. I've attended only one steampunk convention, I confess. The most steampunk costume I've worn was a dress from one of those theme-park stands where you choose an outfit from a rack, dress up, and take family photos for the cost of a set of steak knives. I was about six, and the skirt was almost as wide as I was tall. But, despite my dearth of petticoats, I now live the story that steampunk fans dream of.

I'm a theoretical physicist who works at the intersection of quantum physics, information processing, and thermodynamics. Thermodynamics, the study of energy, developed during the 1800s to explain why steam engines work. Quantum physics—the study of electrons, atoms, and other small particles—underlies a breed of computers that will one day outstrip even supercomputers on certain information-processing tasks. I intersect the futuristic technology of quantum computers with the Victorian setting of thermodynamics, after the fashion of steampunk. This intersection I call *quantum steampunk.* It centers on how quantum physics can transform and enhance thermodynamic technologies, such as engines, as quantum physics is transforming and enhancing computers. Quantum steampunk also provides fundamental revelations, such as a detailed understanding of why time runs only forward, and insights into the rest of the universe, such as black holes. The field has roared to the forefront of scientific advances over the past decade, embodying the excitement and promise of steampunk. This book is intended to impart that excitement and to explore that promise.

We'll begin by meeting the fields that coincide in quantum

steampunk: thermodynamics, quantum physics, and information processing. I'll have to introduce myself, too, as a traveler such as yourself should follow only a known, trusted guide. Then, we'll trace out the route that we'll follow in the rest of the book.

When Shall We Three Meet?

Of the fields that meet in quantum steampunk, thermodynamics is the oldest. *Thermodynamics* is a branch of physics and chemistry that budded during the 1800s—the steampunk era. The steam engine was propelling the Industrial Revolution, powering factories and locomotives. Scientists and engineers wanted to understand how efficiently engines could pump water out of mines. Technology drove thinkers to study which obstacles limit machines, as well as how energy changes form—for instance, how the motional energy of water falling over a dam transmutes into the electrical energy produced by a generator. Fundamental questions emerged from this practical bent: What is heat? Does matter consist of particles too small to see? Why can't we walk backward through time as we can walk backward along Tower Bridge in London?

Thermodynamicists studied the systems they could manipulate and observe: steam, metal, water, and the like. Those systems consist of about 10^{24}—or one with 24 zeroes after it, a billion billion million—particles. We call these systems *classical*. We can predict most of their behaviors using physics codified before the 1920s.

We can't predict the behaviors of quantum systems in the same manner. Quantum systems behave in ways impossible for steamboats, sponge cakes, and spaniels; and quantum systems tend to consist of few particles. Particles are the nuts and bolts of quantum systems; examples of particles include electrons, photons (particles of light), and atoms. Audrey and Caspian can bring two atoms together, and force the atoms to interact, such that the following transpires: Say that the friends bring their atoms together in

London, whereupon Audrey takes one atom to Manchester. Then, Caspian measures some property of his atom. The measurement will disturb his atom and, because both atoms have interacted, will disturb Audrey's atom, too. Audrey's atom can be regarded as changing instantaneously (with some caveats that I won't detail)—even though Audrey's atom is hundreds of miles away. Not that Caspian can send Audrey a message instantaneously, for subtle reasons that we'll explore in chapter 2, but loosely speaking, the friends will have correlated the atoms more strongly than they can correlate any classical systems, or created a relationship between the atoms that's tighter than any relationship achievable between classical objects. We call this relationship *entanglement.*

Quantum theory budded during the 1920s, after its seeds were sown during the late 1800s and the early 1900s. Decades later, scientists shifted from asking how quantum systems work to asking how we can leverage quantum phenomena such as entanglement. *Quantum information* science is the study of how quantum systems can process information in ways forbidden to classical systems. By *process information,* I mean, solve computational problems, such as converting pounds sterling to shillings, communicate information, secure information, and store information in memories.

Imagine a computer that consists not of classical transistors, and not of 10^{24} atoms, but of tens of thousands of atoms. We could entangle the atoms and harness their correlations (we'll see how in chapter 3). Such a quantum computer could solve, in minutes, certain problems that would cost classical computers many years. Potential applications include chemistry, materials science, and drug design. Quantum computers could also break part of the encryption that protects web transactions from hackers. (Don't worry too much; postquantum cryptographers are developing codes that quantum computers can't break.) Quantum computers couldn't help with all our problems; for instance, I wouldn't recommend preparing your taxes on one. Still, commercial giants such as Google, IBM, Honeywell, and Microsoft are building quantum computers. So are startups, such as IonQ and Rigetti, and the governments of multiple countries. Tech giant Amazon offers an online portal through which consumers can use startups' early-generation quantum computers.

Don't expect web encryption to break in the next few years, though. Controlling entanglement among tens of particles has cost generations of graduate students. Controlling entanglement among tens of thousands of atoms will cost loads more. A few skeptics believe that we'll never be able to control entanglement among many particles, although most quantum-computing scientists disagree. Time will tell, provided that funding for quantum computing doesn't dry up first.

To exhibit quantum phenomena such as entanglement, most quantum systems need to operate at low temperatures. Cooling—the expulsion of heat, or random energy—falls in the purview of thermodynamics. But how to measure heat, in quantum contexts, requires thought: measuring a quantum system changes, or *disturbs,* it. In contrast, sticking a thermometer in your (classical) mouth won't influence your fever. But measuring how much heat a few atoms emit can influence the amount of heat emitted. Nineteenth-century thermodynamics needs reenvisioning for twenty-first-century quantum science: we must replace gears, pulleys, and levers in the theory of thermodynamics.

What mathematical, conceptual, and experimental toolkit should we use? Quantum information science. The promise of quantum technologies galvanized the development of quantum information science as steam engines galvanized the development of thermodynamics. Since the Industrial Revolution, scientists have applied thermodynamics to understand everything from the stars to the origins of life. Over the past three decades, scientists have been applying quantum information science to understand anew computer science, mathematics, chemistry, materials, and more.

Quantum information science offers a toolkit for revolutionizing thermodynamics to describe small, quantum, and information-processing systems.

I am participating in the revolution. I'm not the first in the field; nor am I the only revolutionary. My cohorts span the globe, and augurs of our mission whispered as early as the 1930s. Many call our field *quantum thermodynamics* or *quantum-information thermodynamics.* But thermodynamics developed as science was emerging from natural philosophy, which has an aesthetic. Natural philosophers understood aesthetics, as they studied philosophy, literature, and history, in addition to geometry and astronomy. Today's physicists invoke aesthetics, too—when favoring a simple equation that describes much of the world over a complicated equation, or over an equation that describes little. But aesthetics arguably used to play a broader role in science and natural philosophy. For instance, scientific instruments shared the elegance of musical instruments during the Victorian era: brass gleamed against mahogany; a curve occasionally arched without necessity because it pleased the eye. Honoring aesthetics ties one to the grandness, richness, and inspiration of the tradition inherited by today's scientists. Quantum thermodynamics has the aesthetic of steampunk, I realized while pursuing my PhD: I marry thermodynamics with quantum information science—the Victorian era with futuristic technology. Hence the term *quantum steampunk.*

{ Around the World in a Quantum Career }

I first encountered steampunk without recognizing the genre for what it was. When I was in elementary school, on weekend mornings, my family would snuggle into my parents' bed and watch *The Adventures of Brisco County, Jr.* Bruce Campbell starred in the television series, which pits a gentleman cowboy against a time traveler. When I was in fifth grade, Will Smith played a sheriff in the steampunk film *Wild Wild West.* Across fifth grade and middle school, I devoured two steampunk book series: *The Chronicles of Chrestomanci* by Diana Wynne Jones and *His Dark Materials* by Philip Pullman.

As Pullman's series spans many worlds in a multiverse, my journey into physics spanned many disciplines. I wanted to be a Renaissance woman: In high school, I reveled in calculus and mechanics, magical realism and European history. A philosophy teacher imparted a fascination with quantum theory. This fascination drove me to the philosophy department in college, but more kindred spirits surfaced in the physics department. Faculty there helped me construct a major called Physics Modified: I took prerequisites in physics and mathematics, chose upper-level physics courses, and sampled philosophy, history, and mathematics related to physics.

Within physics, quantum computation drew me because of its balance: not only are quantum technologies useful, but they also inform our understanding of space, time, and information—of fundamentals. Even better, the subject crosses disciplines: a quantum-computing physicist wields mathematics; interfaces with computer science and chemistry; knows her history (for example, what Albert Einstein said to his colleague Niels Bohr about measurement); writes papers and presents lectures; and undertakes something—the physical interpretation of equations—that physicists like to think of as unique to physics (although the practice reminds me of the critical analysis of literature). The program convinced me to burrow into theoretical physics, with an eye on quantum computation.

After college, a research assistantship in England led to a master's program at the Perimeter Institute for Theoretical Physics. Perimeter attracts researchers from across the world to Waterloo (near Toronto), a city that excels in physics, mathematics, tech startups, and winter. Waterloo's public library contained a slim novel by Canadian poet Jay Ruzesky. One chapter featured a

French inventor who built automata, or humanlike robots. In one scene, the inventor gazes down upon his workshop, his cloak billowing behind him. Shortly after I read that chapter, a piston slid into place in my mind. I realized that I'd encountered the steampunk genre.

I encountered quantum-information thermodynamics at Perimeter, undertaking my first research in the field there. My PhD took me to Pasadena, California, whose associations with little old ladies is the butt of a joke by Brisco County, Jr., in one episode. Pasadena is home to the California Institute of Technology, or Caltech. Caltech nurtured quantum computation before the subject commanded respect as a field, largely due to John Preskill. John exudes the gravitas you might expect of a Princeton and Harvard alumnus who's as decorated in physics as a general is in the military. But he occasionally sings and dances onstage (poorly, as he'd admit), and a grin that he struggles to control sometimes cracks through his poker face. John supervised my PhD, granting support and independence that I likely will never be able to repay. When we met, I told him that I wanted to work at the intersection of quantum computation and thermodynamics. *Ok,* he said. *Do it.*

I did. I worked morning, noon, and night, occasionally escaping to Vroman's Bookstore in Pasadena. To reach the children's section from the fiction, one climbs a staircase that pauses, midway, at a landing. On the landing, I once discovered drawings of a green-haired girl, wearing a corset and a flouncy dress. Brian Kesinger, a Disney cartoonist, had created the drawings. He depicted the girl sharing her adventures with an octopus, the mascot of steampunk.* Steampunk echoed just a half-hour walk from Caltech. The Huntington Library, Art Museum, and Botanical Gardens boasted

* Octopodes feature in *Twenty Thousand Leagues under the Sea,* and they exhibit intelligence and engineering prowess. For example, an octopus escaped from the National Aquarium of New Zealand, apparently by slithering through a gap in his tank, propelling himself across the floor, and clambering into a drainpipe that fed into a nearby bay.[1]

one of my favorite museum exhibitions: "Beautiful Science" trumpeted the exquisite in astronomy, medicine, and optics. I'd linger over the seventeenth-century manuscripts, the hand-written letter to an astronomer, the drawing of a unicorn amid Renaissance zoology.

The Huntington reflected a facet of science that I missed at Caltech. I adored Caltech—its intellectual boldness; its scholars, who shared ideas and advice; and the friends I found who lacked artifice. I couldn't have wished for more inspiration there: as I pursued my PhD, four Caltech members won Nobel Prizes, and another laureate taught the chemistry course I took. Caltech scientists didn't just work at the cutting edge; one had created one of the world's sharpest edges, by arraying single atoms in a line.

I reveled in Caltech's science, but I hearkened back to natural philosophy. I was helping build the future, yet I yearned for the past. Another piston slid into place in my mind, and I coined the term quantum steampunk to describe my research.

My research involves, apart from helping develop a modern theory of thermodynamics, harnessing the theory to transform other spheres of science. Quantum physics, information, and energy surface across science: in materials science, chemistry, black holes, optics (the study of light), and more. Colleagues and I are wielding our theory to answer questions, and to discover new questions, in these fields.

Quantum Steampunk served as the title of my PhD thesis. Although completing a PhD pleased me, using the word *steampunk* in scientific literature pleased me more. I write this book as a postdoctoral fellow at Harvard University and the Smithsonian Institution's Institute for Theoretical Atomic, Molecular, and Optical Physics. By the time the book is published, I'll be building a research group as a physicist at the National Institute of Standards and Technology (NIST) and the University of Maryland. The institutions have, between them, two quantum institutes, a quantum-technology center, and an interdisciplinary institute that highlights thermodynamics. I'll be adding a quantum-steampunk laboratory for theoretical physicists, who discover facets of the physical world—such as entanglement—by applying and interpreting mathematics. We'll also collaborate with experimentalists to test our predictions.

I'm not one of the steampunkers who appear if you google *steampunk photos*: I don't own goggles or a greatcoat, and I don't frequent steampunk conventions. I've read hundreds more physics papers than steampunk novels. I don't fancy lacing up a corset, although I wear skirts most days. I *have* assembled a curiosity cabinet, containing a spyglass, a butterfly specimen, museum-exhibition posters, and old-fashioned keys collected from Spain and Oxford and Santa Barbara. But I take more pride in the curiosity cabinet that I've constructed by doing physics. This book unlocks that curiosity cabinet, the steampunk life I live.

A LEGEND FOR THE MAP

In chapters 1 through 4, we'll review background needed to approach quantum steampunk: information theory (the study of information processing), quantum physics, quantum computation, and thermodynamics. Chapters 5 and 6 unite these fields, introducing quantum steampunk. I envision part of quantum steampunk as a map of the vintage drawn by explorers centuries ago: ink on parchment, with a dragon or a mermaid in one corner. City-states, kingdoms, and principalities dot the map. They represent different subfields of quantum steampunk—different philosophies, aims, toolkits, and collections of results. We'll visit many of the city-states in chapters 7–13. In chapter 14, we'll step off the map. We'll use quantum thermodynamics as a kaleidoscope through which to view, and from which to glean insights into, the rest of science. The back of the book contains a glossary, in case you'd like to refresh your memory of terms we encounter.

This book celebrates the thermodynamics developed during the 1800s, as well as the steampunk spirit of adventure and exploration. But one nation's adventure sometimes turned into the exploitation of another nation. The Victorian era featured phenomena not to be celebrated, including colonialism, racism, inhumane working conditions, child labor, and environmental destruction. This book's celebration of steampunk doesn't amount to an endorsement of everything Victorian. But quantum steampunk offers an opportunity to revive the nineteenth century's successes while improving on its shortfalls.

Each chapter begins with a snippet from the quantum-steampunk novel that I've dreamt up. The snippets don't represent science accurately, but science does lurk behind the snippets. So let them tickle your fancy, and you can believe the rest of each chapter. Not that I'd have believed the rest before becoming a quantum steampunker.

CHAPTER 1

INFORMATION THEORY

OF PASSWORDS AND PROBABILITIES

A slat of wood rattled as it slid across the grating in the oaken door.

"Password," rasped a voice through the grating.

"You know me perfectly well, Baxter," Audrey snapped. "And I refuse to recite that ludicrous—"

"T'aint Baxter," said the voice, sounding less raspy and more put-upon. "Baxter left for th' water closet 'alf an hour ago."

"Oh. Very well." Audrey drew a breath and dashed through the password as though it were a scandal sheet that she didn't want to be caught reading. "Lord-Buntiford-drank-two-bottles-of-wine-and-then-danced-a-jig-without-clothes-on."

The slat of wood slid back into place with a rattle, and the door creaked open.

WHAT DID AUDREY transmit to the gatekeeper? Her vocal cords energized air molecules, which buzzed around, knocking into other air molecules, which knocked into other air molecules, until the knocking reached the gatekeeper's ears. Inside his ears, air molecules bumped against tiny bones, causing them to vibrate. The vibration stimulated a neuron to fire, which stimulated more neurons to fire, and so on until he opened the door.

Did Audrey transmit energy to the gatekeeper? A neural-firing pattern? She did, but those phenomena don't form the essence of what she communicated: information. This chapter focuses on information—what it is, how we encode it, and how we measure it. The measurement of information will lead us to a concept—entropy—that also stars in thermodynamics.

We live in the information age, but what is information? According to one viewpoint, information is the capacity to distinguish among alternatives. Before Audrey spoke the password, the gatekeeper couldn't tell whether she deserved admittance. She gave him the ability to distinguish between her belonging or not belonging.

I encountered this view of information in a textbook when I first studied quantum computation in college.[1] I've mostly grown into the perspective. Still, I'd tweak one facet of the definition: information doesn't suffice for one to be able to distinguish between alternatives. If you carve the *Encyclopedia Britannica* onto an ice floe, you give the floe information. But the floe won't be able to distinguish a nectarine from Napoleon, as can humans who read the corresponding *Encyclopedia* entries. So, I'd call information an ingredient necessary for distinguishing between alternatives.

The second definition of *information* I learned at Oxford, which I visited while pursuing my PhD. Amid the soaring spires and daily downpours, I met philosopher of physics Chris Timpson. He explained his view of information over lunch at a faculty dining hall, which entailed what an American student might hope for from an Oxford dining hall: linen napkins; a quiet atmosphere; and an upholstered, couch-littered sitting room for taking a postprandial cuppa tea. Chris regards information as that which can catalyze a process without losing its ability to catalyze that process. Say that, in Audrey's story, the gatekeeper drops the door's bolt on his foot. He cries out, and Baxter runs to the rescue from the water closet. Baxter shepherds his colleague to a couch, then returns to the door. He knows nothing of Audrey's presence, but she can knock again. The password, relating that Audrey belongs to the cabal, will recatalyze the opening of the door.

Recall from the prologue that Ewart is the villain who tried to eavesdrop on Audrey using a mechanical spy-fly. What if he learns the password and then the cabal discovers that he's learned it? The group will change the password, and "Lord Buntiford drank two bottles of wine and then danced a jig without clothes on" will no longer cause the door to open. Will information have lost its ability to catalyze a process?

No, the information is "I belong to the cabal." The cabal has encoded that information in the password. If Ewart learns the password, the code ceases to reflect the information reliably. The cabal will junk its code and pick a new password—say, "Lord Buntiford ate three roast suckling pigs at Christmas." If Audrey provides the new password, she'll provide the same information as before: "I belong to the cabal." That information will catalyze the door to open.

The password encodes one message ("I belong") in another message ("Lord Buntiford drank two bottles of wine and then danced a jig without clothes on"). Messages are abstractions. We store them in, and convey them through, physical systems: vocal cords, air molecules, ear bones, neurotransmitters, pixels on a computer screen, and more. We translate an abstract message into a physical system's configuration using another code, such as the English language and alphabet. I encode information (say, our heroine's name) by arranging pixels in a certain configuration on my computer screen (by typing). You decode that information by reading ("Audrey").

So, we encode one message, such as "I belong to the cabal," in another message, such as a password, by encoding one abstraction in another. To store and transmit information, we encode messages in physical systems. The ink in a book illustrates such an encoding.

Teaspoons of Information

We measure things in terms of units—time in seconds, sugar in teaspoons, and length in meters or inches or (if you're webcomic artist Randall Munroe) giraffes.[2] How do we measure information, and what is the unit of information? We'll approach these questions like physicists: start with examples, form a guess, and then check the guess against more examples and against principles, modifying our guess when necessary.

In Audrey's story, the gatekeeper must distinguish whether Audrey belongs to his cabal. Suppose that, before hearing the password, he hasn't the foggiest idea whether she does. From Audrey's communication of the password, he learns a substantial amount of information. Now, suppose that the gatekeeper recognizes Audrey's voice the first time she speaks. He ascribes a high probability—say, 75%—to the possibility that Audrey belongs to his cabal. Hearing the password doesn't surprise him; the password conveys little information.

Events—the hearing of a password, the reading of a book, a lover's getting down on one knee—convey information. The more expected the event, or the higher the event's probability, the less information the event conveys. As an event's probability increases, the probability's inverse decreases. As a reminder, the inverse of two is one-half, the inverse of three is one-third, and so on. Let's guess that, if an event with some probability of happening occurs, the probability's inverse measures the information conveyed.

Let's check how reasonable this guess is. Our rule for measuring information should reflect how amounts of information add up. Suppose that, after Audrey responds, the gatekeeper asks whether the pub next door has closed. The conversation will give the gatekeeper two pieces of information: Audrey belongs to the cabal (rather than not belonging), and the pub is open (rather than closed). The total amount of information learned should be the first amount plus the second.

We can measure the total amount of information another way. The gatekeeper's conversation with Audrey produces one of four possible outcomes: (1) Audrey belongs to the cabal, and the pub remains open; (2) Audrey belongs to the cabal, and the pub is

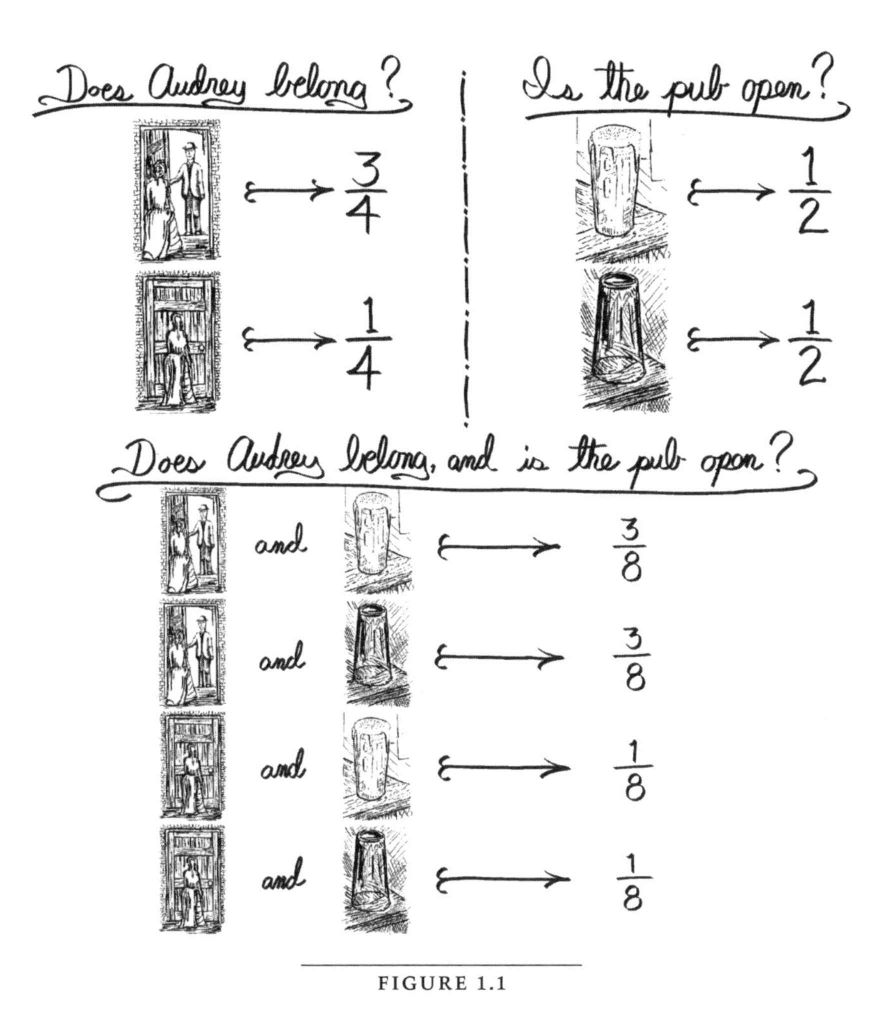

FIGURE 1.1

closed; (3) Audrey doesn't belong to the cabal, and the pub remains open; or (4) Audrey doesn't belong to the cabal, and the pub is closed. Each joint event consists of two constituent events. Each joint event's probability is the first constituent event's probability times the second constituent event's probability. For example, say that the gatekeeper recognizes Audrey's voice but has no idea whether the pub is open. He ascribes a three-quarters probability to Audrey's belonging, a one-quarter probability to Audrey's not belonging, a one-half probability to the pub's remaining open, and a one-half probability to the pub's being closed. The joint event "Audrey belongs, and the pub remains open" has a probability of three-quarters times one-half, or three-eighths (figure 1.1).

How much information does the gatekeeper gain from the joint event? According to our earlier stab at quantifying information, the amount of information is the inverse of the joint event's probability—the inverse of three-eighths, or eight-thirds. But we also concluded that the total amount of information should be the sum of the constituent amounts. Drat, as Audrey would say. We have two expressions for the total amount of information learned by the gatekeeper. One expression is a product, and the other is a sum. We'll have to tweak our rule for quantifying information—to turn the product into a sum, so that both expressions are the same.

The Hungarian mathematician Alfréd Rényi said, "A mathematician is a machine for turning coffee into theorems," or proven mathematical facts.[3] A *logarithm* is a mathematical machine for turning products into sums. By "mathematical machine," I mean that the logarithm takes in numbers and outputs possibly different numbers. For our purposes, turning products into sums is the logarithm's crown-jewel property.

The logarithm will improve our rule for measuring the information conveyed by an event: Let the amount of information be the logarithm of the inverse of the event's probability. This number is called the event's *surprisal* because it measures how much the event surprises you. The more an event surprises you, the more information you learn. According to our revised rule, amounts of information add together, even though event probabilities multiply.

We'd like to break an amount of information down into units, as we can break a bowl of sugar down into teaspoons. What should we designate as the teaspoon of information?

Information, we established, is an ingredient needed to distinguish between alternatives. The least possible number of alternatives is two: Audrey belongs to the cabal or doesn't. Without prior information (lacking familiarity with Audrey's voice), the gatekeeper assigns both possibilities equal likelihoods, of one-half. According to our rule for measuring information, the amount of information he learns is the logarithm of two. This amount is the unit of information, called the *bit*. You learn a bit of information upon flipping a fair coin and (after crawling under the table where it rolled) finding that it landed heads-up.

Imagine an unfair coin that has a three-quarters probability of landing heads-up and a one-quarter probability of landing heads-down. (Equivalently, suppose that the guard recognizes Audrey's voice.) Seeing the coin land heads-up, you learn log(4/3) bits of information. Seeing the coin land heads-down, you learn log(4) bits. Any time a random event happens to you—the weather is sunny rather than snowy or cloudy, or your favorite pub closes early, or your steam-powered time machine breaks down—you can measure in bits the information you've learned.

Bit refers to the unit of information, but information scientists also use the word in other ways. *Bit* can refer to an event that can play out in two ways. In one of the least romantic examples imaginable, the response to a marriage proposal constitutes a bit. Also, we sometimes designate as a bit a physical system able to be in one of two possible states. Examples include a candle that's lit or unlit. The two options—for the physical system or for the event—may be "lit" or "unlit," or "yes" or "no," or "shepherd's pie" or "ploughman's lunch," or any other dichotomy. Information scientists represent the options with zero and one (0 and 1). This convention simplifies our work, much as I'd enjoy reading papers about shepherd's pies and ploughman's lunches. Alas . . .

{ The Liver of Information Theory }

What information is and how we measure it fall under the auspices of information theory. The scientist Claude E. Shannon invented information theory during the first half of the twentieth century. He published the theory while working at Bell Labs, founded by telephone inventor Alexander Graham Bell. Scientists of a certain vintage reminisce about the mid-twentieth-century Bell Labs much as siblings reminisce about summer visits to Grandma's house. The research institution fostered curiosity and collaboration, garnering nine Nobel Prizes. Shannon lived up to the lab's reputation. He wondered how efficiently you can send information down a channel, or communication medium, such as a telephone wire. Communication is an information-processing task, as are storing information, securing information, and solving computational problems. Information theory is the study of how we can quantify information and perform information-processing tasks.

So how efficiently can we send information down a wire? I call the answer "the liver of information theory," thanks to my ninth-grade biology teacher. The teacher, a firecracker of a Texan woman, advised my class, "If you ever don't know the answer to a test question, write, 'liver.'" The liver fills an absurd number of roles in the human body. I memorized about 10 roles in ninth grade; Johns Hopkins Medicine's website reports over 500.[4] If you don't know the answer to a question on a biology test and you answer "liver," you have a high probability of receiving credit. If someone asks, "With what efficiency can you perform such-and-such an information-processing task?" and you answer, "It depends on an entropy," you have a high probability of receiving credit.

An *entropy* is a function of surprisals—a mathematical machine that takes in surprisals and spits out a number. Mathematicians have built many such machines, so many entropies exist. Different entropies measure the efficiencies with which we can perform different information-processing tasks. We'll see a slew of entropies in chapter 10, but we'll focus on one for now: one of the best-known entropies, named after Shannon.

Imagine that Audrey visits the cabal's headquarters every night at 10 p.m. Baxter is supposed to be guarding the door. He has a probability p_W of being in the water closet, a probability p_P of having snuck off to drink a pint, a probability p_S of having fallen asleep, and a probability p_G of standing guard. The numbers p_W, p_P, p_S, and p_G form a *probability distribution*. On any given night, if Audrey finds Baxter gone to the water closet, she learns a number of bits of information dependent on p_W; if she finds him having snuck off for a pint, she learns a number of bits dependent on p_S; and so on. How much information does Audrey learn on average over many nights? The answer is the average surprisal, or the *Shannon entropy*.

Let's see how the Shannon entropy is the best efficiency with which we can perform an information-processing task. To simplify the situation, let's suppose that Baxter only ever stands guard or sleeps: p_G and p_S are the only nonzero probabilities. Suppose that Audrey manages cabal members, checking that they perform their duties. She records Baxter's performance every night, writing a *G* or an *S* in her journal. She repeats this ritual for years, filling a journal with a random string of letters. She decides to squeeze the string into the smallest possible space, performing the information-processing task of *data compression*.

You might have compressed files on a computer before emailing them to friends or colleagues. How do we compress information? Say that the imaginary novel featuring Audrey gives rise to a long book series. At the end of the final novel, Audrey has collected reports for most of 30 years. As 30 years contain about 11,000 nights, her random string contains about 11,000 letters. If the string consisted of just one letter, the string would be one of two possibilities: *G* or *S*. If the string consisted of two letters, it would be one of four possibilities: *GG, GS, SG,* or *SS*. Appending a letter doubles the number of possible strings. So, by the end of the final novel, Audrey has written one of $2^{11,000}$ possible strings. That's

FIGURE 1.2

an enormous number—far larger than the number of atoms in the observable universe.

But not all those possible strings are likely. Probably, about a fraction p_G of Audrey's letters are *G*'s, and about a fraction p_S are *S*'s. Say that, on any given day, Audrey has a decent probability of catching Baxter in either activity; that is, neither probability lies close to 0 or 1. Also, Baxter's habits aren't completely random; he's more likely to stand guard than to sleep.

Audrey might find Baxter standing guard every night for 30 years. But her likelihood of recording 11,000 *G*'s is so low that we can call that string basically impossible. Most strings are basically impossible, mathematics shows. Audrey has a decent probability of writing only strings of a certain type—the strings in which about a fraction p_G of the letters are *G*'s and about a fraction p_S of the letters are *S*'s. This fact parallels how, if we flip a coin 11,000 times, we expect the coin to land heads-up about half the time and tails-up about half the time.

How does Audrey compress her string? She labels the first not-virtually-impossible string 1, the second not-virtually-impossible string 2, and so on (figure 1.2). Her journal probably contains one of those strings, and she records that string's label. That label is the result of her data compression: Audrey has squeezed 11,000 letters into one label. At least, Audrey has replaced 11,000 letters with one label. I haven't shown that the label contains fewer than 11,000 letters. Let's figure out how small the label is—how many bits Audrey needs to specify the label.

Every time she checks on Baxter, Audrey receives information. How much information—how many bits—on average? The answer is the Shannon entropy of p_G and p_S, which measures her average uncertainty about what she'll find Baxter up to. Across 11,000 days, Audrey receives a number of bits that about equals the Shannon entropy times 11,000. That's how many bits form Audrey's label. Say that Baxter is four times more likely to stand guard than to sleep: p_G is 4/5, and p_S is 1/4. The Shannon entropy turns out to be about 0.7 bits, so the label requires about 0.7 bits for each of the 11,000 days, or about 8,000 bits total. Data compression saves Audrey 11,000 minus 8,000 bits—or 3,000 bits—measured with the Shannon entropy.

YOU WILL ALWAYS HAVE THE ADVANTAGE

What we call "the Shannon entropy," Shannon called "entropy." Why did he choose that word? He explained the reason in a 1971 *Scientific American* article:[5] Shannon had built the average surprisal as a mathematical machine for measuring uncertainty. Uncertain about what to call the machine, he consulted the Hungarian-American mathematician and physicist John von Neumann. (Rényi, who called mathematicians machines for turning coffee into theorems, was another Hungarian mathematician. We'll meet more Hungarian scientists later in this book: studying information theory, quantum physics, and thermodynamics, you can't escape from brilliant Hungarians.) Von Neumann advised Shannon about what to call the average surprisal: "You should call it entropy, for two reasons. In the first place your uncertainty function has been used in [a variation on thermodynamics] under that name, so it already has a name. In the second place, and more important, no one knows what entropy really is, so in a debate you will always have the advantage."

Three elements of von Neumann's advice merit comment. First, we'll confirm a claim of von Neumann's in chapter 4: an entropy does crop up in thermodynamics. Entropies also crop up in quantum physics, as we'll see in chapter 2. Entropies unite information theory, thermodynamics, and quantum theory, as we'll discuss in chapter 5.

Second, von Neumann claims that "no one knows what entropy really is." I expect that he was speaking tongue-in-cheek. Physicists had been shedding light on the notion of entropy for years. Shannon had proved what role entropy plays in data compression, and von Neumann had lent his name to the quantum version of the Shannon entropy. So, von Neumann knew more than he appeared to give himself credit for.

Nevertheless, the concept of entropy has always carried an aura of mystery to me. Earlier in the chapter, we constructed the Shannon entropy from intuitions about how information behaves. Examples showed us why the Shannon entropy should have the form that it has. Yet that form brings to mind a scene in the novel *The Pinhoe Egg* by the twentieth-century British novelist Diana Wynne Jones. A boy finds an enormous egg in an attic, takes the egg home, and cares for it. The egg cracks open one night, revealing a bedraggled brown form that has too much skin on its back and a scraggly tail. Reading the book, I didn't guess that another character would remark, "Bless me, it really is a griffin!" A griffin? The creature looks like a starved Dachshund that's fallen in a lake. It'll soar through the sky one day? Arch its back, raise its wings, and command more majesty than a bald eagle atop a California mountain? I think about entropy similarly: That hodgepodge of components—two copies of a probability, a fraction, and a logarithm—holds the key to information theory? This mathematical abstraction explains the arrow of time (as we'll see in chapter 4), an effect felt in our bones and skin? The tension between the entropy's potpourri form and visceral impact captivates me. So, I agree about the challenge von Neumann alludes to, a challenge in wrapping our minds around entropies. That challenge has bewitched me since I first encountered an entropy, in that ninth-grade biology class.

As for the third claim, von Neumann encourages Shannon to cleave to entropies because, "in a debate you will always have the advantage." What better reason to read this book?

CHAPTER 2

QUANTUM PHYSICS

EVERYTHING AT ONCE, OR, ONE THING AT A TIME?

A painting beckoned to Audrey from across the hallway, from an alcove tucked away from the hum and bustle of the natural philosophers. She approached without hearing her boots tapping on the marble floor; she wove between pairs of murmuring gentlemen without seeing them. The painting depicted a wood-paneled study, the likes of which Audrey yearned to curl up in with a pencil, a notebook, and silence. Against the study's far wall stood a desk formed from three stacked shelves, the top of which formed the writing surface. The desk supported a globe, two mechanical contraptions, and what could have been a mathematical compass or a nutcracker. A broad sheaf of paper rested on the desk, trapped beneath a pile of books, and partially hanging off the wood.

Audrey stepped closer. Was she looking at a desk, or was it a platform onto which the scholar could climb? The platform hypothesis gained support from a mahogany-colored door, behind the globe, that led out of the study. Above the door, drawings had been inked on the wall; Audrey peered at them more closely. Was that a tree inked on the right-hand side? Was light entering a lens inked on the left? Leftward of the drawings stood a column, plainer than its Ionic cousins but evoking the classical world of Euclid and Pythagoras. Leftward of the column, a gold-latticed window afforded a view of the sky.

A sigh escaped Audrey as she peered through the painting's window: in the bottom panes, a cloud drifted above gleaming water that wove through a golden-brown landscape. The study must have been floating in the heavens—and what more fitting place for a heavenly study? Hands clasped behind her back, Audrey bent forward and read the brass label beneath the painting.

"Everything at once, or, one thing at a time?"

PAINTINGS, STEAM ENGINES, submarines, and heroines fall within the purview of classical physics. By *classical physics*, I mean the union of three theories: classical mechanics, electrodynamics, and general relativity. Isaac Newton established *classical mechanics* during the 1600s. This theory describes objects big enough to be seen with the naked eye, or under a classroom microscope, and how they move—why the Wright brothers' plane flew, the effort required to raise an obelisk, and how gears make a clock tick.

James Clerk Maxwell pioneered *electrodynamics* during the 1800s to describe light. Light includes the radiation we can see, such as the glow emitted by the gas lamps that lit London during the steampunk era. Electrodynamics describes also the infrared light we radiate, the ultraviolet light that damages skin during sun exposure, and other radiation invisible to humans.

Electrodynamics inspired Albert Einstein to formulate his theory of *general relativity*, which debuted during World War I. General relativity describes large and massive objects, such as planets and stars. By applying general relativity, we can send spacecraft across the solar system and reason about the universe's shape.

All physical phenomena not described by these three theories, we call *nonclassical*. Quantum physics is nonclassical. Systems described by quantum theory tend to be small and to weigh little. For example, a hydrogen atom measures about one ten-millionth of a millimeter across. That speck of matter weighs 10^{24} times less than the sugar you'd need to bake one shortbread biscuit. Hydrogen atoms, as well as other quantum systems, exhibit behaviors that classical systems don't. This chapter focuses on quantum behaviors. We'll illustrate with the behavior that earned quantum physics its name, *quantization*. Then, we'll overview more behaviors that will crop up as we tour quantum steampunk.

⋄{ MIND THE GAP }⋄

Let's contrast a classical behavior with a quantum one. You could witness the classical behavior on Halloween at my graduate university, Caltech. Every Halloween, Caltech undergraduates tote pumpkins to the roof of the old library. The students freeze the pumpkins with liquid nitrogen, and then toss them off the roof.

Why the old library? It's Caltech's tallest building. The higher the undergrads tote the pumpkins, the more energy the students invest in the pumpkins. The pumpkins acquire *gravitational potential energy*, which comes from defying the Earth's pull. The more gravitational potential energy the pumpkins acquire, the more energy they have to transform into motional energy as they fall. The greater the pumpkins' motional energy when they hit the ground, the greater their speeds, and the greater the liquid-nitrogen-frozen carnage.

A pumpkin can gain gravitational potential energy from ascending the old library's nine stories, or one story, or half a staircase, or one step, or two-seventeenths of an inch. The pumpkin can basically acquire or lose any amount of energy, from no energy to a library's worth.

The energy gained and lost by the pumpkin is gravitational potential energy. An atom can have energy of another type—a type that's common but, according to quantum theory, available only in certain amounts. To describe this energy, let's focus on a hydrogen atom because it's simplest. A hydrogen atom consists of a proton, which forms the nucleus, and an electron. The proton carries a positive electric charge, and the electron carries a negative charge. The attraction between the charges, together with the electron's motion around the nucleus, endows the atom with energy. Of this type of energy, the atom can have only fixed amounts. It's as though the atom can climb or descend Caltech's old library only via a ladder, sketched in figure 2.1. Each rung represents some amount

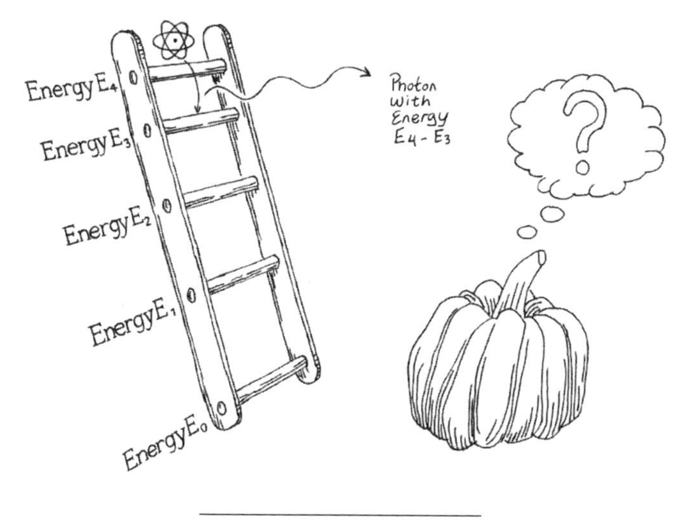

FIGURE 2.1

of electronic energy that the atom can have. The atom can ascend from a lower rung to an upper rung, or descend from an upper rung to a lower. However, the atom can't come to rest between rungs.

For instance, the atom in the figure begins on a high-up rung, with an amount E_4 of energy. The atom can descend to the next-highest rung while emitting a packet of energy—a *photon*, or particle of light. The photon carries the difference between the two rungs' energies. Then, the atom can repeat this process—descend to the next rung while emitting another photon. Each photon carries off an amount of energy calculable with quantum theory. Imagine the atom descending, step by step, to the ladder's bottom rung. The final photon carries 10^{-19} times less energy than the gravitational energy of a pumpkin sitting one yard off the ground.

We call the packets *quanta* of energy because they contain fixed amounts. Likewise, the atom's electric energy is said to be *quantized* because the amounts accessible are fixed. "Quantum physics" means "physics of fixed-size packets." Which makes my work sound like the science of airline snacks; but I've heard graver insults.

In middle school science class, we envision electrons as particles zipping around the nucleus. This vision doesn't capture the whole truth—we'll see how later—but the picture captures some important features of the truth; so it's worth using sometimes. The electron follows a path that curves around the nucleus. The electron's *velocity* consists of the electron's speed and direction; and as the electron curves, its direction changes. Its velocity therefore changes, and so the electron accelerates.

Also as we learn in middle school science class, the electron carries negative charge. Charged particles usually radiate, or emit, photons as they accelerate. This radiation enables us to play Billie Holiday, one of my favorite jazz musicians, over the radio. The radio DJ forces electrons in an antenna to vibrate up and down. Every time the electrons switch direction—for instance, from up to down—they accelerate and emit photons. The photons travel to your receiver, which helps transform their energy into sound.

Curving around the nucleus should cause electrons to radiate photons, which would carry away energy. The more energy the electrons radiated, the less these negatively charged particles could resist the attraction of the positively charged nucleus. The electrons would spiral into the nucleus, and the atom would implode. Matter couldn't exist; radios and DJs and ears couldn't exist; and we'd never hear Billie Holiday's music. Since I have heard Billie Holiday's music, this story, based on classical physics, must be wrong.

Quantization rushes to the rescue by allowing the atom to have only certain amounts of electronic energy. Namely, the atom's energy ladder ends with a lowest rung. When the atom occupies that rung, it can't drop any lower. On that rung, furthermore, the electron orbits the nucleus one ten-millionth of a millimeter away. The electron can't spiral into the nucleus because it can't emit photons, because the atom lacks any lower energy rung to which to drop.

Why does the ladder have a lowest rung? Quantum theory offers little insight about such questions. Physics elucidates *what* and *how* questions. It also sheds some light on *why* questions, such as "Why does the sky look blue?" or "Why don't atoms implode?" or "Why does a ponytail have the shape it has?" (Yes, really.)[1] But why should the laws of physics be such that atoms don't implode? Attempts to answer this question veer into philosophy, circularity, or theology. I endorse the questions but lack the expertise to answer. At least atoms *don't* implode; quantum theory reveals what prevents them from imploding; and matter's stability allows us to keep asking questions.

Quantization isn't as nonclassical as one might think; a classical system's energy can act almost as though it were quantized. For example, consider a jar of apricot preserves owned by Audrey's family, the Stoqhardts, since her father was a boy. The jar has occupied one or another of the pantry's ten shelves for decades. A shelf's height determines the gravitational potential energy that the jar has when on that shelf. So, the jar's gravitational potential energy has been one of ten fixed numbers for most of several decades.* That is, the energy behaves almost as though it were quantized.† How nonclassical a behavior is will concern us later, when we evaluate which parts of quantum steampunk we could mimic classically.

For now, we'll just survey behaviors exhibited by quantum systems. We'll begin with *spin*.

* Granted, the jar's gravitational potential energy changes fluidly whenever the chef moves the jar to a higher or lower shelf. But the jar occupies a shelf throughout most of the years.

† Granted, every object in our everyday world consists of quantum particles and so has quantized energies. But the energy-ladder rungs are separated by a teensy amount because the object is large. So, the distance between the energy rungs—the quantization—is basically impossible to observe. We can regard the ladder rungs as basically touching each other, and we call the object classical. In contrast, the jar's approximate energy quantization is observable and is imposed by the jar's classical environment, not by quantum physics.

You Spin Me Right Round

Imagine a ballerina—clad in a white leotard, her hair in a bun, ribboned pointe shoes on her feet. She shoots straight upward, balances atop one pointe-shoe tip, spins twice, freezes, and returns her raised foot to the ground seconds later—without grimace or wobble, as if the feat cost no more effort than a yawn. The ballerina reigns as queen of angular momentum. *Angular momentum* is a quantity, like energy, possessed by every object rotating about an axis. How much angular momentum depends on the object's mass, its speed, and how far each chunk of it lies from the axis.

An electron in an atom has angular momentum. The electron orbits the nucleus, according to middle school science class. Quantum theory contains a mathematical model for the electron's angular momentum.

But the electron doesn't have just angular momentum; it has another property, called *spin*, described by the same mathematics. Ballerinas don't have spin, despite having angular momentum. Classical systems don't have spin. Quantum particles—electrons, photons, protons, and more—do.

This fact puzzled early quantum physicists. What, they wondered, is this angular-momentum-like thing? Some conjectured that the electron doesn't only orbit the nucleus as the Earth orbits the sun, but also rotates about its axis, as does the Earth. To check their conjecture, physicists gathered measurements of the electron's spin, mass, and length.* They estimated the speed at which the electron would have to rotate to achieve the measured spin. That speed turned out to be greater than the speed of light. As no matter moves more quickly than light, the electron couldn't rotate at that speed. Spin

* The electron doesn't technically have a length, for a reason explained later. But we can, loosely speaking, imagine the electron as filling a certain volume. One coarse estimate of this volume's length is 10^{-13} inches.[2]

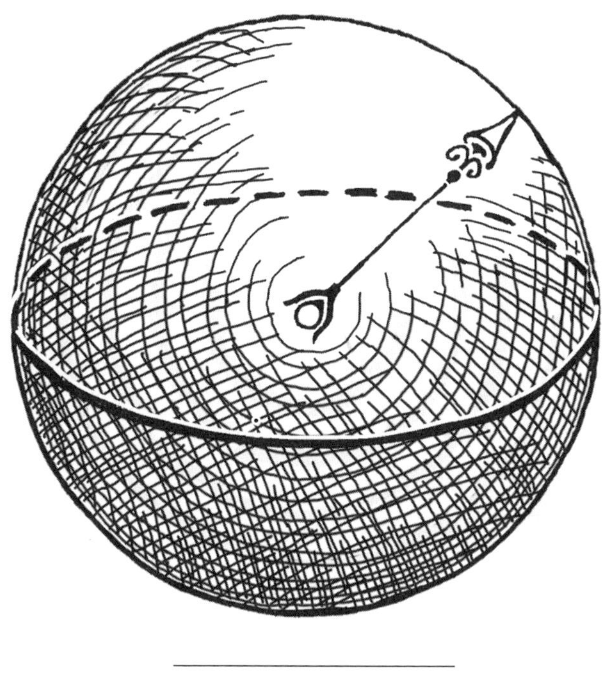

FIGURE 2.2

must not stem from rotation, physicists concluded, despite being described by the same mathematics as angular momentum. But we have one more tool for wrapping our heads around spin.

Velocity consists of a speed—a positive number, a negative number, or zero—and a direction. Angular momentum, too, consists of a number and a direction. The direction points along the axis about which one rotates. For instance, a ballerina's angular momentum points parallel to her spine. We can also attribute a direction to an electron's spin. We can control the spin's direction using a magnet, as you can control the direction in which a screw points using a fridge magnet. I imagine each electron as clutching the base of a teensy arrow that can point in any direction: The arrow's tip can touch any point on a sphere that encloses the electron (figure 2.2). That arrow represents the electron's spin.

FIGURE 2.3

Why care about spin? We'll encounter a reason in the next chapter: spins store information in a simple way. While we can often imagine spins as teensy arrows clutched by electrons that resemble teensy balls, this cartoon sometimes spits on the truth too forcefully. Our next quantum phenomenon explains why.

{ MAKING WAVES }

I prepared for my career in physics, during childhood, by jumping rope. At recess, friends and I took turns turning the rope tied to a metal fence. Occasionally, after the school day ended, I wandered by and experimented with shaking the rope.

Shake one end of a rope, and a wave propagates down it. The distance between successive crests is the *wavelength*. Shaking slowly produces a long wavelength, and shaking quickly produces a short wavelength (figure 2.3).

Waves don't only propagate down ropes, or in the ocean. Quantum theory ascribes wave properties to matter. This ascription partially explains why I don't fully endorse the middle school vision of the electron. Viewing the electron as a minuscule ball zipping around the nucleus, we can understand some properties of the electron. For example, imagine measuring an electron's position. Our detector will flash the coordinates of some location, akin to the coordinates that flash on a GPS screen. But the electron isn't truly a miniature ball sitting at that point. How does the electron differ?

We can ascribe a location to the electron only at the moment when we measure the particle's position. Before the measurement, and shortly after the measurement, the electron resembles a wave more than a ball. A wave doesn't occupy just one point. It extends across a distance, consisting of crests, troughs, and the spaces between. Quantum theory ascribes to the electron a wave that extends throughout space. The wave peaks near an atom's nucleus. The higher the wave rises, at any point, the more likely is our detector to flash that point's coordinates if we measure the electron's position. We can't predict the measurement's outcome, usually, because the wave stretches throughout space. We can predict only the measurement's probability of pointing to this position or that position.

So the mathematics that describes the waves in a jump rope also describes the electron. According to quantum theory, every chunk of matter and light resembles a wave. We call this property *wave-particle duality*, and it extends even to you. Your wavelength depends on your speed, as when you're strolling down the street. While strolling, you have a wavelength 10^{26} times smaller than a hydrogen

FIGURE 2.4

atom: properties of you undulate across that distance as a wave undulates across a jump rope. Don't worry that the undulation will fill your life with tremors, as though you lived in San Francisco and quantum theory were an earthquake. No one can observe such short lengths, so you'll never notice your wavelike properties.

One behavior of waves crops up throughout quantum physics, as well as in a jump rope. Imagine a jump rope held, at opposite ends, by Audrey and Baxter. Audrey shakes her end upward and downward at some rate. A wave, formed from a string of peaks and troughs, will travel from her toward Baxter. Baxter can create a wave, too; for example, he can shake his end of the rope more quickly, without ever lifting his hand as high as Audrey does (figure 2.4). The rope will be in a *superposition*, a sum of waves that's a wave itself. I didn't master superpositions in kindergarten, though I could skip rope through minutes' worth of the rhyme "Cinderella dressed in yella."

Like a jump rope wave, an electron can be in a superposition. Imagine measuring an electron's position and reading the location off our detector. Quantum theory describes the electron with one tall wave pulse—as if Audrey snapped the rope upward and downward once—a pulse that peaks at the location read off our detector. If the detector had yielded another location, the wave would have peaked there. Imagine waiting for a while after our measurement. The peak will drop and broaden. The low, wide result turns out to be a superposition of constituent waves, each peaked tightly at one point. The electron, we say, is in a superposition of positions.

It isn't that the electron occupies one position or another and we don't know which. Rather, the electron doesn't have a well-defined position. In a sense, the electron responds to the painting that entranced Audrey at the beginning of this chapter. "Everything at once," the painting's title asks, "or, one thing at a time?" "Everything at once," the electron responds, in a sense.

The painting that arrested Audrey has a twin in our world—a twin that hangs in the National Academy of Sciences headquarters in Washington, DC. I wandered inside the building the spring after declaring my college major. As the painting mesmerized Audrey, so did it mesmerize me. A miniature of it now sits in my office, courtesy of the painter, Robert van Vranken. My desk doesn't resemble the desk in the painting, but I view the work on my desk as a reflection of the artwork.

A HUNDRED INDECISIONS

We've explored four phenomena critical to quantum theory: quantization, spin, wave-particle duality, and superpositions. The next phenomenon features in a 1915 poem by the American-British poet T. S. Eliot, "The Love Song of J. Alfred Prufrock." The speaker, a middle-aged extra in the play of life, wrestles with himself:

> And indeed there will be time
> To wonder, "Do I dare?" and, "Do I dare?"
> Time to turn back and descend the stair.

As uncertainty haunts the speaker, so does it haunt the quantum physicist.

To see how, let's expand our electron discussion from position to momentum. *Momentum* reflects the difficulty of stopping an object: the heavier the object, and the more quickly it moves, the greater its momentum. Suppose that we've just measured an electron's position. The position now has a well-defined value, according to our earlier discussion. But the electron's momentum doesn't have a well-defined value. If we now measure the momentum, our detector can report any speed and any direction. Furthermore, all possible momenta are equally probable. Our uncertainty about the momentum couldn't be greater.

Why? If the electron has a well-defined position, then it's in a superposition of different momenta. The converse is true, too: having a well-defined momentum, an electron is in a superposition of positions.

We've been focusing on extremes—certainty about position and maximal uncertainty about momentum and vice versa. But quantum physics accommodates intermediates. Say that we've trapped an electron in a box, so that we can pinpoint the particle's location to within a few nanometers. If we measure the electron's momentum, we'll obtain one of just several possible numbers; not all momenta are possible anymore. The more certain the position is, the more uncertain the momentum, and vice versa. This property is called the *uncertainty principle*.

The uncertainty principle extends beyond position and momentum. Recall that an electron in an atom, orbiting the nucleus, has angular momentum. The electron also has an angular position. Imagine gazing down at the atom as at an analog clock you've laid on a table. The electron can be hovering above the 12, or beside the 3, and so on. Uncertainty about the angular position trades off with uncertainty about the angular momentum. Catch the electron below the 6—at a well-defined angular position—and the electron lacks a well-defined angular momentum: the particle zips around the clock without any particular speed.

If you've read other quantum physics books and articles, you might have seen the name Heisenberg connected with quantum uncertainty. Werner Heisenberg, a German physicist, cofounded quantum theory during the early twentieth century. He latched onto uncertainty like a quantum J. Alfred Prufrock. Heisenberg intuited the relationship between position and momentum, estimating the trade-off between them.[3] Later physicists, beginning with Earle Hesse Kennard,[4] crystallized Heisenberg's realization in mathematics, in an *uncertainty relation*.

No wonder Heisenberg stars in jokes bandied about in physics departments. For example, one day, a police officer stops Werner Heisenberg on a highway. "Do you know how fast you were driving?" the officer demands. "No," Heisenberg admits, "but I know where I am."

We encountered uncertainty outside of quantum theory in chapter 1: Audrey didn't know whether she'd find her brother guarding the door, in the water closet, nursing a pint, or sleeping. We measured her uncertainty with the Shannon entropy. One might therefore expect to measure quantum uncertainty with entropies. Can we? Yes—as J. Alfred Prufrock, in Eliot's poem, measures his life in coffee spoons.

Imagine an electron in any state—in one location, or in a superposition of all locations, or in a superposition of a few locations. Imagine measuring the electron's position. Our detector has a probability p_1 of reporting this location, a probability p_2 of reporting that location, and so on. Those probabilities have a Shannon entropy, like the probabilities with which Audrey finds Baxter engaged in this or that activity.

Now, imagine that we measured the momentum instead of the position. Our detector had a probability q_1 of reporting this momentum, a probability q_2 of reporting that momentum, and so on. These probabilities, too, have a Shannon entropy. The two entropies—the position entropy and the momentum entropy—can't sum to a very small number, according to the uncertainty principle. The lesser one entropy, the greater the other.

So, we can't escape uncertainty in quantum physics, as Prufrock can't in his poem. Prufrock's uncertainty extends beyond himself to his interactions with others, as he asks, "Do I dare / Disturb the universe?" He wouldn't have the chance to bite his nails over the question if he were a quantum experimentalist. All experiments involve measurements, and measuring a quantum system disturbs it.

Classical systems lack such sensitivity. Imagine wandering along a beach, finding a piece of driftwood, and examining it in the sunlight. The light hits the wood, which absorbs some of the photons. The other photons bounce off and enter your eyes. Those photons don't harm the wood, which is fortified by its large size and many particles. The driftwood can confront many photons without, so to speak, blinking an eyelash.*

Quantum systems blink like a teenager who's stumbled into the kitchen, wearing pajamas, hair sticking out at all angles, at noon. In the previous section, we imagined an electron with a well-defined momentum—a particle moving with some speed in some direction. Such an electron is in a superposition of locations. We imagined measuring the electron's location, as by shooting photons at the electron and observing where they end up. From a photon's final position, we infer where the photon crashed into the electron, changing course. The electron lacks the wood's heft, the wood's indifference to a few photons. So, the crash traumatizes the electron, making the momentum completely uncertain. The

* Granted, light harms wood in extreme situations. For example, by concentrating light on wood through a magnifying glass, you can start a fire. But driftwood can withstand many hours of sunlight without altering much.

measurement has disturbed the particle. In other words, measurement disturbance distinguishes quantum theory from classical.

You may have heard that, in quantum physics, a measurement "collapses a wave function." Collapsing a wave function amounts to turning a superposition into one wave that has just one peak. We collapse a wave function by preparing an electron in a superposition of locations, then measuring the particle's location, forcing the electron into one spot. I won't use the term *wave function* much, as quantum-computing scientists don't; I'll talk about measurement disturbance instead.

And They Lived Entangled Ever After

I married during the COVID-19 pandemic, while a postdoc in the Boston area. My mother had been planning a wedding near my childhood home. But, not wanting to cause the deaths of my relatives, my elementary school teachers, and some of the world's leading quantum researchers, I cancelled the arrangements. Massachusetts had just poked its head out from its lockdown shell. So, my fiancé and I swore our fidelity in masks, on an empty patch of Harvard Yard, before twelve guests and a bottle of sanitizing wipes.

But humanity can't resist uniting and celebrating, even amidst a pandemic. Sixty-odd guests attended the ceremony via Zoom. Congratulations poured in via email, social media, and greeting cards. A bouquet, ordered by a colleague of mine, surprised my husband and me on our doorstep. The card tucked into the bouquet said, "May you stay forever entangled." Another card we received read, "I'm really happy at your entanglement."

What accounts for this wedding custom of quantum physicists? *Entanglement* is a relationship that quantum particles can share. The relationship leads to correlations stronger than any creatable with classical particles.

Two systems are *correlated* if changes in one system track changes in the other system. Audrey's parents, the archaeologists Drs. Lewis and Loretta Stoqhardt, can illustrate correlations for us. The parents are on a dig in Iraq, and each parent writes the children a letter every day. The first paragraph relates the parents' love, how much they look forward to seeing the children, etc. The second paragraph's subject varies. Suppose that both parents write in the evening, after discussing the day's events. If Audrey's mother writes about the gypsum reliefs found that week, Audrey's father likely will. The father will likely switch to describing a statue on the day that the mother does. The subjects will be highly correlated.

Now, suppose that the mother writes at night, while the father writes in the mornings. The pottery sherds that absorb the father on Wednesday morning might occupy the mother that night. But the Arabic poetry that she hears between morning and night might distract her. The paragraphs' subjects will be less correlated.

Now, suppose that Audrey's mother is creating a dictionary of the Akkadian language. Carvings, translations, and vocabulary permeate her thoughts and dreams. The topics fill the second paragraph of each of her letters, while the father's second paragraphs drift from carvings to the desert to indigestion. Her subject will be uncorrelated with his.

Understanding correlations, we can approach entanglement. Imagine Audrey and Baxter bringing two quantum particles, such as electrons, close together. The particles might thereupon interact. For example, two close-together electrons repel each other because both carry negative charges. Some interactions create entanglement, which can manifest as follows.

Say that the siblings entangle their electrons' spins, putting the spins into a superposition. Let Baxter take his particle far from Audrey's—to the opposite side of the room, or of the city, or even of the globe. Let Audrey measure any property of her spin—for example, whether the spin points upward (figure 2.5a). She'll have

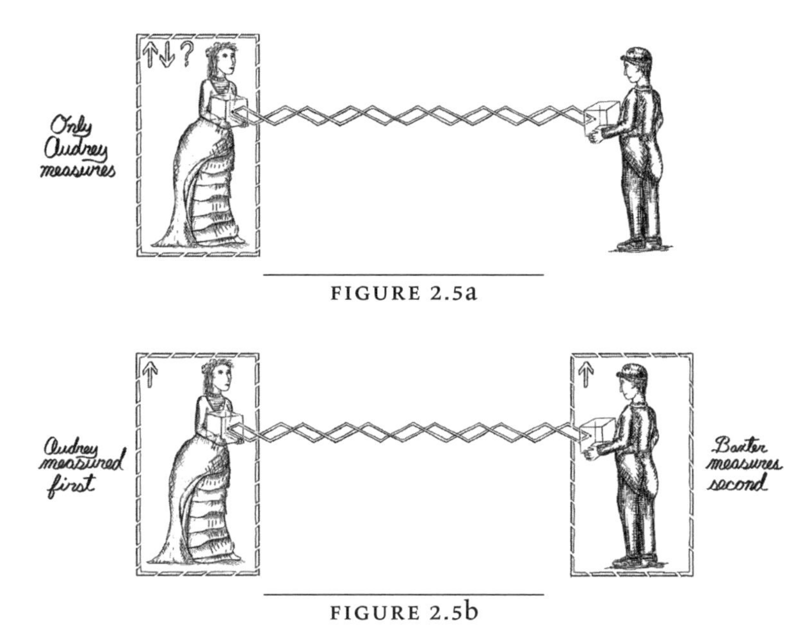

FIGURE 2.5a

FIGURE 2.5b

no idea whether her measurement device will read "yes" or "no." But, when Audrey measures her spin, she disturbs it, inducing it to point upward or downward only. Thanks to the entanglement, Audrey's measurement also disturbs Baxter's spin: Suppose that Baxter measures his spin as Audrey measured hers. Audrey will be able to predict his outcome with certainty (figure 2.5b), although she couldn't have before her measurement because the spins were in a superposition.* Loosely speaking, Audrey's measurement of her particle influenced Baxter's particle, although Baxter was far away, due to the entanglement.

The foregoing story may give us pause, but entanglement

* In figure 2.5b, I'm assuming that Audrey and Baxter share an entangled state of a particular type, in addition to assuming that Audrey obtains the "yes" outcome. The nature of this entangled state ensures that if Audrey obtains a "yes," then Baxter obtains the same outcome, rather than the opposite.

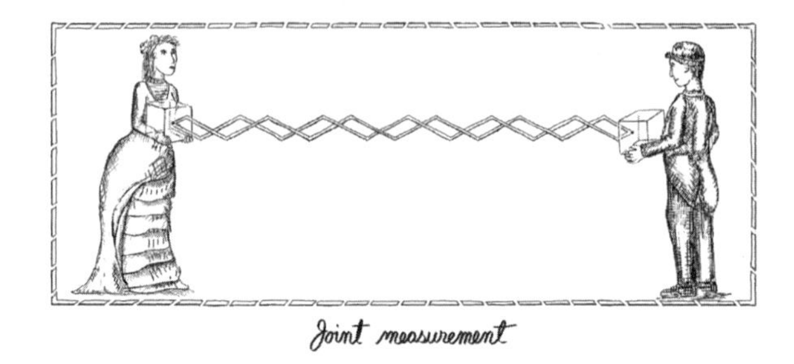

FIGURE 2.5C

challenges our intuitions further. Suppose that, after entangling the spins, the siblings don't measure their particles individually. Instead, the siblings perform a special joint measurement of both the spins together (figure 2.5c). Audrey and Baxter can predict the outcome with certainty before performing the measurement.

This conclusion should surprise us. Audrey can predict nothing about her particle; no matter how she measures it, she has no idea what her detector will report. That is, Audrey has no information about the outcome of any measurement of her particle. Neither does Baxter have information about the outcome of any measurement of his particle. But the siblings have complete information about the outcome of a measurement of both particles; the siblings can predict the outcome perfectly. So, the siblings have complete information about the pair of particles, yet no information about either individual particle—complete information about the whole, yet no information about the parts.

In classical physics, if we know everything about the whole, we know everything about the parts. For example, before my wedding, I knew that a Michelle and a Miles would attend; I knew everything relevant about the whole pair of guests. Therefore, by classical-physics logic, I knew everything relevant about the parts: I knew that Michelle (one part) would attend, and I knew that Miles (the other part) would attend. So, Audrey and Baxter's situation, in which they know everything about the whole but nothing about the parts, sounds like codswallop. But a whole entangled system is greater than the sum of its parts.

I think of entanglement as something shared between particles. The entanglement isn't in one particle, and it isn't in the other. It isn't in the sum of particles measured individually. It's in the collection of particles.

This collectiveness enables entanglement to produce correlations stronger than any producible by classical particles. We've imagined Audrey and Baxter measuring whether their particles' spins point upward. In figure 2.5b, Baxter's detector reads "yes" if and only if Audrey's does; the results are correlated perfectly. This correlation sounds strong, but classical particles can mimic them. We can imagine that, when Audrey and Baxter brought their particles together, the particles flipped a coin and agreed to respond "yes" if the coin landed heads-up and to respond "no" otherwise. (Not that particles can flip coins or speak. But they can behave in ways that have the same effect.)

So perfect correlations in such simple experiments shouldn't impress us. But we can design more-devious experiments, as the physicist John Stewart Bell did in 1964.[5] Bell requires Audrey to choose her measurement (to choose which property she measures) randomly in each trial and Baxter to choose *his* measurement randomly in each trial. The details of Bell's experiment lie outside the scope of this book; you can find them in John Gribbin's book *Schrödinger's Kittens and the Search for Reality*.[6] We need to know only the following: Suppose that Audrey and Baxter run many trials of Bell's experiment. Audrey obtains many measurement outcomes, as does Baxter. The siblings can calculate the correlations between their measurement outcomes—how much the changes in Baxter's outcomes track the changes in Audrey's outcomes. The correlations can be stronger than any producible with classical particles.

Hence my colleagues' wedding wishes. "May you stay forever

entangled" meant, "May you share a strong partnership." My husband and I weigh far more than electrons, take up far more space, and consist of far more particles; therefore, classical physics describes us, and we can't entangle. But isn't the thought sweet?

{ What You Didn't Learn in Driving School }

Now that we've encountered entanglement, let's delve into some of its traits. Let's return to the experiment in which the siblings entangle their particles, then Audrey measures hers, then Baxter measures his. Before Audrey measures, Baxter has no idea which outcome he'll receive. But, after hearing Audrey's report, he can predict his outcome with certainty. Why?

To answer that question, we'll have to extend our understanding of superpositions. We've established that a particle can be in a superposition of locations or a superposition of momenta. Also, a spin can be in a superposition of pointing in different directions. Imagine a wave stretching across a qubit sphere, with one peak at the north pole and one peak at the south. Such a spin fails to point in a well-defined direction. You can measure whether the spin points upward or downward, causing a disturbance that induces the spin to point in one direction. Furthermore, a set of spins can be in a superposition of all pointing in one direction and all pointing in another direction. We can't draw that wave on a two-dimensional page, but experimentalists can prepare such a superposition in the laboratory.

Audrey and Baxter, when entangling their particles, can form a superposition of two upward-pointing spins and two downward-pointing spins. When Audrey measures her spin, she disturbs it, inducing it to point along one direction. Her measurement, with the entanglement, induces Baxter's spin to point along the same direction (figure 2.5b).

This inducing happens instantaneously, even if Baxter transported his spin across the country after creating the entanglement with Audrey. Not even light can traverse a country instantaneously; light travels only at about 10^9 miles per hour. According to Einstein's theory of relativity, nothing can move more quickly than light. Does quantum theory contradict relativity?

No, as we can see by recasting the problem in terms of information. Relativity dictates that no *message decipherable by Baxter* can reach him before light can. Say that Baxter measures the same property, with his spin, that Audrey measured with her spin. For example, each sibling measures whether his or her spin points upward. Baxter will obtain the same outcome as Audrey. He'll know the direction in which her spin points, which we can regard as a message from her. But Baxter can extract that message only if he knows which property to measure. Audrey could have measured whether her spin points upward, or leftward, or backward, or slightly upward and slightly forward but mostly to the right. Audrey must communicate her choice of property, whether by telegraph or mail or fire signal or homing pigeon. A homing pigeon takes longer to cross the country than light does.*

So, entanglement doesn't violate Einstein's theory of relativity: Audrey effectively sends Baxter a message in a bottle more quickly than light can travel. But the bottle remains sealed until after light can arrive. Our inability to send messages more quickly than light can travel is called the *principle of no signaling*. No signaling is one rule of physics that won't please your driving instructor.

We've seen one way in which Audrey and Baxter can entangle

* Why can't Audrey tell Baxter her choice of measurement before the siblings separate? Warning Baxter would amount to cheating: Audrey is hoping to communicate information (her measurement's outcome) to Baxter ultra-quickly, using entanglement. If she communicates her choice of measurement before the game begins, Baxter starts with some of the information needed to infer Audrey's measurement outcome. Figuring out Audrey's outcome is a little like figuring out a crossword puzzle. Knowing Audrey's measurement at the outset is like figuring out a crossword puzzle that's already filled in partially—it's cheating.

particles' spins—so that, if both siblings measure whether their spins point upward, identical outcomes will result. The outcomes are maximally correlated, and we say that the particles are *maximally entangled*. Instead of entangling his spin maximally with Audrey's, Baxter can entangle his spin maximally with a spin of Caspian's. But Baxter can't entangle his spin maximally with Caspian's spin *and* maximally with Audrey's; entanglement is *monogamous*. As my colleagues enjoyed reminding my husband-to-be.

Not all entanglement is maximal, however. Baxter can entangle his spin partially with Caspian's and partially with Audrey's. If Audrey and Baxter measure whether their spins point upward, the siblings will obtain the same outcome only sometimes; the outcomes will share weaker correlations. So, we can generalize our definition of monogamy: the more Baxter's spin entangles with one system, the less it can entangle with the rest of the world.

Baxter is a boy of the sort whose pockets remain filled with lint but have enough holes to lose every shilling he receives. One shouldn't trust Baxter to shield his particle from stray photons, nearby atoms, and other threats. His particle will rapidly entangle with everything in sight. The new entanglement dilutes the desirable entanglement shared with Audrey and Caspian. The environment, we say, *decoheres* Baxter's spin. No measurements of the three spins can produce strong correlations. In this way, decoherence downgrades quantum systems, causing them to resemble classical ones. Hence the full message on the wedding card tucked into the bouquet: "May you stay forever entangled, with no decoherence."

{ LET US GO THEN }

We've overviewed seven quantum phenomena: *Quantization* limits an atom to having only certain amounts of energy. *Spin* obeys the same mathematics as angular momentum but doesn't stem from rotations. We can picture an electron's spin as an arrow pointing in some direction. *Wave-particle duality* likens quantum systems to waves spread across space. As you can superpose waves on a jump rope, so can a quantum system occupy a *superposition* of locations (or momenta, or spin directions, and so on). If you measure the system's location, you can't predict which spot your detector will report. The better you can predict a position measurement's outcome, the less you can predict a momentum measurement's outcome, and vice versa, by the *uncertainty principle.* Measuring a quantum system *disturbs* it, forcing a superposition across locations into just one location.

Particles can share *entanglement*, information encoded not in one particle or in the sum of particles addressed individually, but in the collection of particles. Entanglement is monogamous. The more entanglement Baxter's particle shares with Audrey's, the less entanglement Baxter's particle can share with the rest of the world. The latter entanglement decoheres particles, preventing them from producing strong correlations.

Let us go then, as J. Alfred Prufrock says, you and I. Let us go from learning about quantum physics to putting it to work in information processing.

CHAPTER 3

QUANTUM COMPUTATION

EVERYTHING AT ONCE

Audrey whispered the painting's name to herself again: "Everything at once, or, one thing at a time?" She remained tilted forward for a moment, head inclined toward the painting's brass label, hands clenched behind her back, before straightening up suddenly, like a compressed spring uncoiling.

"Everything at once," she declared.

HAVING OVERVIEWED information theory and quantum physics, we'll combine them in a brief survey of quantum computing. Quantum computers leverage quantum phenomena, including entanglement, to process information more efficiently than classical computers can. As quantum computers store information differently than classical computers, we'll begin with the basic unit of quantum information, analogous to the classical bit. We'll then pinpoint why quantum computers are useful and how they came to capture industries' and governments' hearts and purses. We'll see what a quantum computer is made of and then meet our first quantum liver—I mean, quantum entropy.

Caspian often drops by the Stoqhardt home, looking for Audrey. The servants, treating him as a member of the family, show him in. Caspian may find Audrey in the library, reading or writing at a mahogany desk. If he doesn't find her, he flips one of the pencils in her pencil cup upward, leaving the rest pointing downward. The pencil relates one bit of information to Audrey when she returns: Caspian visited (if a pencil stands point-upward) or didn't visit (if all pencils stand point-downward).

The amount of information needed to distinguish between two equally likely alternatives is a *bit*, as we established in chapter 1. We can represent a bit with a physical system that's in one of two possible configurations: a pencil can point upward or downward; a neuron can fire or remain silent; current can flow or not flow through a transistor; a shilling can land heads-up or heads-down.

An electron's spin represents a quantum bit, or *qubit*. Certain other quantum objects represent qubits, too, just as pencils, neurons, and coins all represent bits. But the electron's spin offers a simple example. Although defined as the unit of quantum information, *qubit* sometimes refers to a physical thing that encodes a unit of quantum information, such as a spin.

We represent the electron's spin with an arrow, similar to a pencil. But Audrey's pencil cup constrains her pencil to point upward or downward. The electron's spin can point also leftward; rightward; forward; backward; forward and a hair downward; rightward, a shade backward, and halfway upward; and in any other direction, as we saw in chapter 2. Infinitely many directions exist. The pencil's two directions represent the two alternatives between which Audrey can distinguish upon seeing the pencil. The more alternatives Audrey can distinguish among, the more information she gains. So, can we gain an infinite amount of information from a qubit?

Our answer is the one that Audrey gives Baxter almost any time he proposes one of his schemes: no. Quantum physics restricts us to measuring only one two-outcome property of a spin. We can measure whether the spin points upward or downward, or leftward or rightward, or along a different axis. Worse, our measurement disturbs the spin, as we saw in chapter 2. Disturbing the spin changes the information it encodes, just as thumping on a table spread with Scrabble tiles—making the tiles jump and reorder themselves—changes the information encoded in the tile arrangement. If the spin initially points rightward, and we measure whether it points upward or downward, the spin will end up pointing upward or downward. So, measuring multiple properties sequentially won't reveal the direction the spin originally pointed in.

So why store information in qubits, if measuring a qubit reveals only one bit of information? Why not store that bit in a pencil's pointing upward or downward? We won't corrupt a pencil by observing its direction as we corrupt any qubit we measure. Furthermore, we can devise a stand that lets the pencil point in any direction. Audrey can learn more than one bit from such a pencil. Pointing upward can mean that Caspian came and went; pointing leftward can mean that he's waiting in the billiards room; pointing half-upward-and-half-rightward can mean that he'll return in an hour; etc. The pencil seems able to encode more information than can the electron spin. So why store information in qubits?

Because qubits can entangle, and entanglement can help us solve problems with fewer qubits than bits—exponentially fewer qubits in some cases. We can understand why with help from the Sunday newspaper. When I was in middle school, I adored the puzzles in the Sunday paper. Someone in my family would bake croissants or banana bread; my parents and brother would claim other sections of the paper; and I'd solve riddles and hunt for words. The Jumble puzzle presented me with an unintelligible string of letters and a row of empty squares. I had to unscramble the string into a word, which I'd write out by filling each square with one letter.

The simplest strategy that's guaranteed to work, information theorists call *brute force*. We form every possible permutation of the letters until the word appears. Forming every possible permutation would fill many sets of squares. For example, say that we're given the string *HIGAFRNT.* We'll pick one of the eight letters to fill the first square, then one of the seven remaining letters to fill the

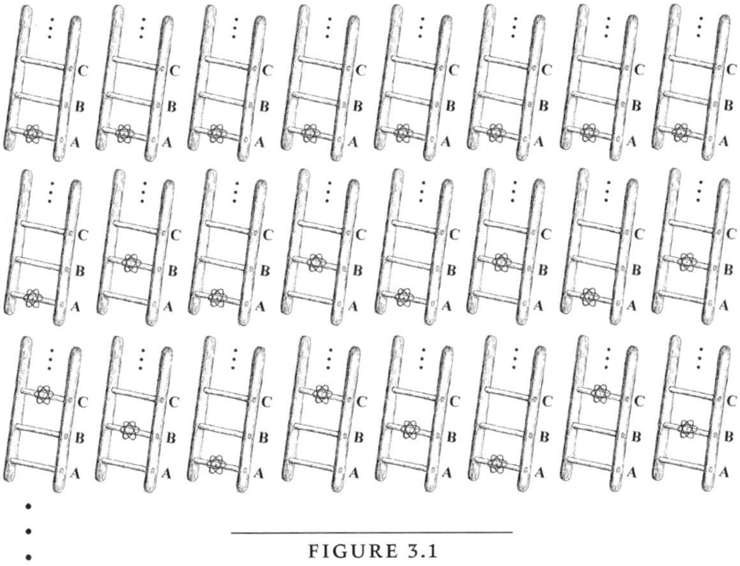

FIGURE 3.1

second square, and so on. We'll need about 40,000 sets of squares to form all the permutations.

Suppose that, every Sunday, our Jumble contains one more letter than the previous Jumble. The amount of paper we'll need to brute-force-solve the puzzle will grow each week. The growth would be huge—comparable to exponential growth. I'd welcome a quantum alternative to the growing mountain of squares.

Let's quit representing each letter with an arrangement of pencil lead on paper. Instead, let's represent each letter with a rung in an atom's energy ladder. Occupying its lowest rung, an atom encodes an *A*; occupying its second-lowest rung, a *B*; and so on for the rest of the 26 lowest rungs. (We can typically prevent the atom from climbing higher—from acquiring more energy—by keeping the atom cool.) We can take eight atoms and put them in a superposition of the relevant energies—a superposition of the possible letter orderings. We can even superpose all the orderings of any eight letters, as shown in figure 3.1.

We don't need hordes of squares to encode all the letter orderings; eight atoms suffice, thanks to superpositions. Quantum systems can, in a sense, store information more compactly than classical systems can. Furthermore, preparing a superposition can take less than a second; whereas writing out 40,000 letter orderings would take days. More generally, quantum systems can help us solve certain problems much more quickly than classical computers can.

Despite these advantages, we can't solve the Jumble puzzle just by putting the atoms in a superposition. Only one component of the superposition—only one of the rows in figure 3.1—represents the puzzle's solution. We have to extract the solution from the atoms, by running some *algorithm*—by following some recipe terminated with a measurement. The simplest algorithm would require no steps before the measurement. Immediately after preparing the superposition, we'd measure every atom's energy, obtaining some permutation of letters. But that permutation would be a random selection from all the possible permutations—and likely not the solution to our puzzle. Extracting solutions from superpositions requires cunning. Broadly speaking, we must prune the wrong components from the superposition as much as possible.

Quantum computer scientists specialize in such pruning, like gardeners armed with quantum physics instead of with shears. Quantum computer science centers on a task that generalizes the solution of a Jumble puzzle: using quantum computers to solve computational problems more quickly than classical computers can. Quantum computers consist of atoms or other quantum objects, rather than today's transistors.

How do we typically solve a computational problem using a quantum computer? We begin similarly to how we'd begin solving a problem in high school math class. There, we'd pull out a blank sheet of paper, which we'd later fill with scratch work. Like our high school selves, a computer needs blank paper, at least

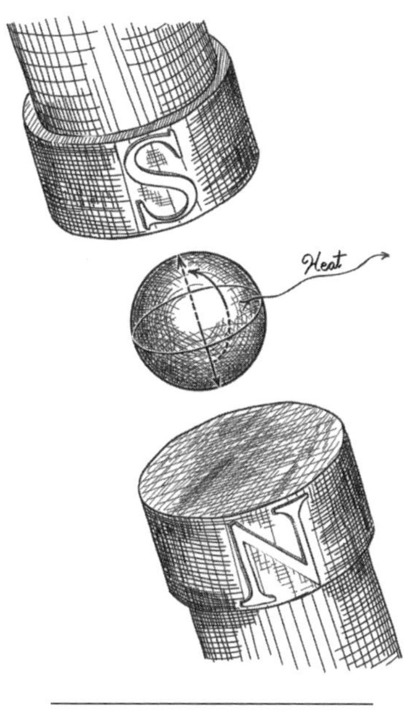

FIGURE 3.2

metaphorically. In a classical computer, "blank paper" consists of transistors that encode bits set to 0. The classical computer's computation flips some bits to 1, then flips other bits, then maybe flips some bits back to 0, and so on. The bits' final configuration records the answer to the problem being solved. A quantum computer has, rather than transistors, electron spins—or atoms or other quantum objects—that encode qubits. The qubits start out in the quantum analog of 0—pointing upward. How do we induce qubits—say, electron spins—to point upward? We can stick a magnet's south pole above the electrons and stick a magnet's north pole below them (figure 3.2). The magnetic field points from north to south, or bottom to top. Then, we cool the spins, lowering their energies. A spin has the least energy when aligned with the magnetic field, so the spins end pointing upward.

After preparing the qubits, we run an algorithm, applying quantum versions of logic gates. Logic gates are the basic steps in a computation, similar in spirit to "plus" and "minus." During the algorithm, we create a giant superposition. Then, loosely speaking, we prune away the superposition's undesirable components as much as possible. How to prune is unclear for many computational problems, as manipulating a superposition is trickier than shaping a shrub. Quantum computer scientists have developed algorithms for solving only a few computational problems with quantum computers. The algorithms leverage entanglement.

At the end of the algorithm, we measure whether each qubit points upward. If we could prune all the undesirable components in our superposition, the solution to our computational problem would result. But we typically can't prune the unwanted components entirely. So, our measurement has only a high probability of outputting the solution. To mitigate this shortcoming, we "rinse and repeat": prepare the qubits and run the algorithm again. And again. After rinsing and repeating enough for a shampoo commercial, we take the majority vote; we guess that the most common measurement outcome represents the solution. If we run enough trials, our probability of guessing correctly can be as high as we please.

Quantum computers can solve certain computational problems much more quickly than classical computers can—in some cases, exponentially more quickly. For other problems, quantum computers offer no advantage over their classical cousins. For instance, I wouldn't recommend that you balance your checkbook using a quantum computer. The Jumble puzzle is an in-between case: a quantum computer can solve the puzzle somewhat more quickly than a classical computer can, but not necessarily exponentially more quickly.* Still, *quantum speedups* have galvanized

* Why not necessarily exponentially more quickly? Because, loosely speaking, of the difficulty of pruning undesirable components from a superposition.

the rise of quantum computing across research, industry, and government.

Peter Shor discovered one of the first, most famous, quantum speedups. Peter is a mathematician at the Massachusetts Institute of Technology (MIT), and he exudes the air of a bespectacled cherub. When visiting MIT, I often encounter him smiling mildly, a ring of white curls framing his face.

Peter sparked the quantum-computing craze that has mushroomed to consume governments, companies, and universities. He lit the spark in 1994, by presenting an algorithm for factoring numbers into primes. A *prime number* is divisible only by itself and one. Examples include 2, 3, 5, 7, and 11.* You may have factored numbers in middle school math class. For example, 75 has the prime factors 3, 5, and 5.

How could a middle school problem ignite a field of science? We can't factor all numbers as easily as we factor 75. For instance, would you have guessed that the prime factors of 879,337 are 719 and 1,223? You could have identified these factors by listing all the prime numbers less than 879,337; dividing 879,337 by the least prime; checking whether the result is a whole number; if not, dividing 879,337 by the next-least prime; and so on. We know of quicker algorithms, but they all take a long time if the given number is large.

Peter invented an algorithm for factoring numbers quickly on a quantum computer. The larger the number, the more the quantum algorithm resembles Speedy Gonzales beside our classical algorithms. The gap grows almost exponentially as the number grows.

* According to a joke among mathematicians, three experts in different fields have to identify a pattern in the prime numbers. The physicist says, "Three is prime, five is prime, and seven is prime. So, the odd numbers must be prime." The biologist says, "Three is prime, five is prime, seven is prime, and eleven is prime. The odd numbers must be prime, while nine must have a genetic anomaly." The engineer says, "Three is prime, five is prime, seven is prime, nine is prime, eleven is prime . . ."

Your inner middle schooler is applauding, I'm sure. But why should governments care? Because prime factoring underlies cryptography that secures many World Wide Web transactions. Imagine wanting to check your bank account online. You open a web browser, navigate to the bank's page, input your username and password, and pull up your account statement. Only you and the bank should be able to see the statement. So, the bank *encrypts* the statement before sending the statement to your computer. Encryption makes the statement look, to eavesdroppers, like a random string of symbols. Your computer decodes the encrypted message, enabling you to verify that you spent $100 at Miss Froo-Froo's Pet-Pampering Salon (on a gift for your poodle-obsessed aunt, of course).

We encode and decode the message using a method called *RSA encryption*. *RSA* stands for *Rivest-Shamir-Adleman*, formed from the names of three MIT scientists. Their encryption method relies on the difficulty of factoring large numbers. If anyone could factor large numbers quickly, they could hack web transactions. Governments are racing to build the first quantum computer, to prevent other countries from breaking RSA encryption before they do. Don't fret about your next order from Miss Froo-Froo's, though. Years—and likely decades—will pass before quantum computers are large enough and reliable enough to threaten your purchases.

Build a Quantum Computer, and the World Will Beat a Path to Your Lab

Most people attribute the notion of the quantum computer to Richard Feynman. Feynman won a Nobel Prize in 1965 for elucidating quantum properties of light and how light interacts with matter. He devised simple ways to solve hard problems in quantum theory. For example, imagine that we're chemists designing a molecule, or a material that exhibits quantum properties. We have to predict how the material will behave in heat, in cold, in magnetic fields, and so on. We generate predictions by simulating the material on a computer. Imagine simulating just a tiny snippet of a material, then a larger chunk, and then the whole material. As the object simulated grows, the time required grows exponentially. A quantum computer, Feynman realized, could simulate a quantum material using only about as many particles as that material contains.

Feynman proposed the notion of a quantum computer in 1982.[1] Yuri Manin proposed the notion two years earlier.[2] Manin is a Russian mathematician who works in Germany. I learned about his insight while pursuing my PhD and working as a teaching assistant in a course taught by Alexei Kitaev. Alexei codesigned a type of quantum computer based on topology, a field of mathematics related to geometry. Alexei lived in Russia years ago, and he highlighted his Russian colleague in the course. Manin's insight exemplifies a mantra repeated by Westerners in certain subfields of physics: no matter what you discover, a Soviet journal probably published a paper about it between the 1960s and 1980s.

Conceived by Manin and by Feynman, the idea of a quantum computer incubated during the 1980s. Paul Benioff theorized how you could build a computer from quantum constituents. Thermodynamics motivated him, I'm delighted to report. Scientists had been arguing about whether all computers waste energy. Does my laptop heat up because all computers are doomed to, or can a computer avoid spilling its battery's energy into the air? To ascertain how little energy could leak, Benioff turned to the physics of the little:[3] quantum theory.*

A few years later, David Deutsch delineated the computations

* Computers can avoid wasting energy, scientists concluded—but only by operating infinitely slowly.[4] I prefer my laptop, regardless of overheating, to a computer that would put me off till the universe ends to tot up my monthly grocery expenses.

that a quantum computer could and couldn't perform. David is a University of Oxford physicist who cultivates a reputation for hermitism. He rarely leaves his suburb, so I used to visit him at teatime while collaborating with physicists in Britain. Sometimes, I'd ask a question—about physics, philosophy, religion, education, fairy tales, grammar, novels, or Pokémon—and David would offer to "pontificate." I found that his pontifications echoed with depth and defied orthodoxy. Such thinking enabled him to lay the groundwork for quantum computing.

Feynman had proposed a *simulator*, a special-purpose computer that solves problems of only one type. You'd build a simulator to study one class of materials or one class of molecules. Other computers are *universal*; you can program them to solve any solvable computational problem, then reprogram them to solve any other. David generalized Feynman's quantum simulator to a universal quantum computer. His vision is one we encountered earlier: a set of quantum particles undergoing quantum logic gates. So, David amalgamated quantum physics with computer science. Scientists and engineers are now implementing his vision, as well as Feynman's and Manin's.

David began "pontificating" about quantum computers in a 1985 paper.[5] I can't resist highlighting one point in that work. David compares a principle of computer science with a law of thermodynamics, attributing the same logical status to both. Computation can't escape thermodynamics.

Nine years after David's landmark paper—in 1994—Peter Shor dropped his thunderbolt. His factoring algorithm inspired physicists, computer scientists, mathematicians, and electrical engineers to switch fields. Not that the field of quantum computation existed yet. I imagine the 1990s as the "cowboy days" of quantum computing. The community was small and scrappy and had little in the way of conferences, funding, respect, or job opportunities. The first students raised in the quantum Wild West completed their PhDs in the late 1990s and early 2000s. Experimentalists had begun implementing small quantum circuits by then. Momentum gathered until departments, governments, and companies could no longer ignore quantum computation.

Funding poured into the field—which gained recognition as a field—around the time I pursued my PhD, during the 2010s. Google, IBM, Microsoft, Honeywell, and other companies hired quantum physicists as though quantum physicists were software engineers. Startups dotted the tech landscape as saloons dotted the Wild West. Universities across the world clamored to establish quantum institutes. Quantum computation traded its chaps and spurs for slacks and loafers.

Over the past few years, popular culture has caught quantum fever. The television show *Devs* spotlights a software engineer who works for a quantum-computing company. United Parcel Service (UPS) calls its package-tracking service Quantum View. A dishwasher detergent is called Finish Quantum; and an antiperspirant is called Sure for Men Quantum Deodorant.* Articles about quantum computing appear in the *New York Times*, the *Guardian*, and the *Washington Post*. As the villain says in the 2018 film *Ant-Man and the Wasp*, "You can forget nanotech. Forget [artificial intelligence]. Forget cryptocurrency. Quantum energy is the future. It's the next gold rush. I want in." The film's sequel—expected to debut within a year of this book—is subtitled *Quantumania*.

My perspective on the frenzy aligns with many other quantum scientists'. We appreciate the world's enthusiasm for our work. We share that enthusiasm (if we didn't, we wouldn't dedicate so much of our life's blood to research). We're grateful for the investments that fund our science, our training of students, and our development of technologies. But we feel obliged to throw cold water on the hype.

* Albert Einstein called entanglement "spooky action at a distance." My husband calls the deodorant "spooky olfaction at a distance."

Quantum computers today contain up to 50-ish qubits apiece. Errors beset these computers as ants beset an abandoned picnic. Realizing quantum computing's potential will require tens or hundreds of thousands of qubits. Most quantum physicists regard this potential as possible to achieve.* But the achievement requires time, patience, toil, and funding that doesn't dry up like a lemonade bottle at a picnic. Scientists joke that quantum computers are 20 years away because they're *always* 20 years away. I expect that we'll clinch quantum computing, but not for many years. Nor will quantum computers solve all problems more quickly than classical computers. Neither do I expect to buy a quantum computer for my home as I'd buy a laptop. (Granted, decades ago, experts didn't expect the public to want personal classical computers.) Let's brew a cup of chamomile tea, take a deep breath, and focus on the truth beyond the hype.

QUANTUM-INFORMATION STATE OF MIND

Quantum computers will impact information security someday, if not tomorrow. So, cryptographers are searching for codes resistant to quantum computation. On the flip side, quantum physics offers resources to cryptographers. For example, recall measurement disturbance from chapter 2. Touching a quantum-encrypted message disturbs it. Consequently, senders and receivers can detect disturbances by eavesdroppers. Institutions such as banks are deploying quantum cryptography to protect data. Banks don't need quantum cryptography yet. But they will someday, many scientists expect, and the early bird gets the secure credit card.

Many scientists expect quantum computers to bear fruit in the form of simulations first. Simulators, having less powers than universal computers, require fewer resources. Quantum simulators

* A minority dissents.

have already engendered discoveries, such as features of phases of matter.[6] Ice, water, and steam exemplify the three phases that we encounter every day: solid, liquid, and gas. Many-particle quantum systems can occupy other phases. Entanglement distinguishes some of these phases as rigidity distinguishes solids from liquids. Quantum simulators can enter these phases, which may guide us toward designing new materials.

But I value quantum computers for more than their applications. The construction of quantum computers has engendered technologies applicable outside of computing. Examples surface in metrology, the study of measurements—of fundamental limits on the precision with which we can measure anything, of which measurement strategies yield the most information, and so on. Entanglement and uncertainty have boosted the precision with which we measure length: Near the end of my PhD studies, two Caltech physicists received a Nobel Prize for the detection of gravitational waves. Gravitational waves are squeezings and stretchings of the fabric of space. The waves detected emanated from black holes that collided far away, long ago. Upon reaching Earth, the waves squeezed and stretched space here through minuscule distances. Scientists detected those tiny squeezes using quantum metrology.

Another example of quantum metrology embarrassed me when I visited the University of Colorado Boulder two years ago. That university, like the University of Maryland, shares a quantum institute with the National Institute of Standards and Technology (NIST). The Colorado institute runs the world's most precise clock, based on the quantization of atoms' energies. More than a day into my visit, I realized that I hadn't set my analog wristwatch back two hours after arriving from the East Coast. A digital wall clock, labeled *NIST*, seemed to glare at me with red numbers that resembled eyes.

Quantum computing has engendered not only technologies

that you can touch and see, but also mathematical and conceptual tools. Scientists now analyze quantum systems in terms of the information that those systems store, transform, and transmit. To illuminate what distinguishes classical physics from quantum physics, we identify information-processing tasks that quantum systems can perform and that classical systems can't. We view the quantum world through the lens ground by Claude Shannon, the founder of information theory. We are quantum information scientists.

Quantum information science spans computation, cryptography, communication, and metrology. It stretches from physics and computer science to chemistry, mathematics, electrical engineering, biology, and materials science. Within physics, quantum information theory has impacted atomic physics, optics (the study of light), particle physics, nuclear physics, condensed matter (the study of solids and liquids), biophysics, black-hole physics, and quantum gravity (the attempt to unite quantum theory—the physics of the small—with general relativity—the physics of the large). Nor, we shall see, has thermodynamics escaped quantum information's influence.

Seth Lloyd captured the quantum-information state of mind in his book *Programming the Universe*. Seth is an MIT professor about whom I have a guideline: If you dream up any idea about quantum computation, he probably mentioned it in a paper a few decades ago. Seth wrote that all particles encode information. Whenever particles collide, they undergo a logic gate. Not necessarily a logic gate that'll help you compute how much to tip your taxi driver. But we can learn about the physics of particles by tracking how their information transforms. We can regard every process—every collision, every interaction—as information processing, or computation. I'll use the terms *quantum information science* and *quantum computation* nearly interchangeably, although technical differences distinguish them.

·{ LAB TOUR }·

What do quantum computers consist of? Most of today's best computers are classical and consist of transistors.* Granted, quantum theory describes how electrons flow through transistors. But transistors serve the same purpose as the vacuum tubes used in computers during the early 1900s. Transistors represent bits as vacuum tubes do—just more compactly. The particles in my laptop can't entangle with each other (much, or usefully). Entanglement underlies quantum speedups, so quantum computers need quantum alternatives to transistors. Experimentalists and engineers are building quantum computers from many *platforms*, or types of hardware. I'll overview three here, and we'll encounter others later in this book.

One type of hardware, I first encountered at the IBM research facility an hour's drive from New York City. Boasting sweeping architecture frosted with glass and stone, the facility reminded me of Fred Astaire: decades-old, yet classy. The technology inside is anything but old. IBM is building a quantum computer from *superconducting qubits*, tiny circuits cooled to low temperatures. *Superconductivity* is a property that graces certain quantum materials: current can flow through the material forever, without dissipating. Imagine current flowing counterclockwise in a superconducting circuit. The current plays the role of an upward-pointing electron spin, acting as the quantum analog of a 0 bit. A clockwise current plays the role of a downward-pointing spin, or a quantum 1. The current can also be in a superposition of flowing in both directions.

When I visited IBM's quantum-computing lab, it contained

* Until a couple of years ago, I'd say, "Classical physics describes today's computers." But quantum computers exist now. They're small and riddled with defects, like the last apple left at the farmer's stand at the end of market day. But quantum computers exist.

seven canisters the size of linen closets. Experimentalist Nick Bronn gave me a lab tour. Upon finding a canister that wasn't running, he climbed half-inside. Gold- and silver-colored wires, trays, and tubes surrounded him. A cinematographer couldn't have conjured a more steampunk scene in Hollywood.*

"This is the fridge," Nick said.

A romantic would hope that quantum computers, a technology of the future, would look futuristic. A dreamer would envision the canister, the silver, and the gold as the quantum computer. But the canister does to the quantum computer what a Frigidaire does to a salmon filet. The wires help the experimentalists wrangle the qubits into computing. And the quantum computer? It consists of a chip that fits in your palm. Granted, the chip is shiny. And designing it required the innovation and toil of a steampunk invention.

I jest about the fridge, but it deserves as much applause as the chip does. And no one should applaud louder than a quantum thermodynamicist. Cooling—expelling heat—is a thermodynamic process. Nick's fridge cools superconducting qubits to near absolute zero, the lowest temperature attainable. Upon learning how far such fridges cool their qubit filets, my husband expressed astonishment at their name.

"*Fridges*?" he said. "They cool to the lowest temperatures in the world, and physicists couldn't at least call them *freezers*?" You can call them *dilution refrigerators*, if you want to sound more scientific.

The second quantum-computing architecture nearly impacted my entanglement—I mean, my wedding plans. A year after I moved to Harvard, my then-boyfriend asked what type of engagement

* Others had recognized the technology's steampunk air, I learned later. Three-and-a-half years after visiting IBM, I presented about quantum steampunk at the Yale Quantum Institute. Faculty member Rob Schoelkopf and others are building another quantum computer from superconducting qubits. Rob had learned the term *steampunk*, he told me, from a journalist who'd visited his lab.

ring I'd like. I joked that I'd prefer a diamond riddled with defects because I could run quantum algorithms on it. Diamond consists of carbon atoms arranged in a repeating pattern. Imagine expelling two neighboring carbon atoms and replacing one with a nitrogen atom. The resulting structure has electrons that can encode qubits. A jeweler would call such areas *defects*: they discolor the diamond, lowering its desirability as a decoration. Moreover, quantum-computing diamonds are either nanoscale—too small to see—or rectangular plates mounted on a special material. My boyfriend gave me an heirloom stone instead.

The third quantum-computing platform consists of nuclei in atoms. Nuclei have spins, similarly to electrons, that can serve as qubits. Many nuclei—and therefore many qubits—cluster together in a molecule. We can control these qubits with a magnetic field, to perform quantum logic gates. This control, combined with measurements of the spins, forms an experimental toolkit called *nuclear magnetic resonance* (NMR). Medical doctors use NMR to image people's brains in magnetic resonance imaging (MRI) scanners. Not that your doctor runs quantum computations on your brain. But MRI uses magnetic fields to identify nuclear spins in your brain and take a picture of them. Like your brain, a quantum computer run on NMR can operate at room temperature. Not needing fridges the size of linen closets offers an advantage. But scaling up an NMR quantum computer—stuffing a molecule with nuclei—is difficult chemically.

We've glimpsed three platforms for quantum computers: superconducting circuits, defects in diamond, and NMR. Many more platforms exist, and different companies and universities are betting on different favorites. Each platform, like a breed of horse, offers pros and cons. Comparing the contenders would require another book, so I won't do so here. Which platform will win the race to quantum computing? No one knows. And hybridizing platforms—leveraging each to accomplish what it excels at—might triumph. I'm cheering for the hybrids, as an interdisciplinarian less partial to the *c*-word *competition* than to the *c*-word *collaboration*.

How to Insult a Quantum Information Theorist

As a theorist, I appreciate quantum computers but prefer the mathematics that describes what they do. That mathematics is called *linear algebra*. What does it consist of? When I studied the topic in college, I tried to explain to a friend what it was about. Remember the simplest equations you encountered in middle school? They'd describe how many puppies Janie walks, if Janie and Johnny walk eleven puppies total and Janie walks three more puppies than Johnny. Linear algebra is the study of those equations. Perhaps I should have lied, since my friend responded, "For this, you had to go to college?" Yes, I did. In middle school, we had to solve two of those equations at a time. Quantum computing requires us to solve thousands. But if you ever want to insult a quantum information theorist, say, "Pffft. Isn't quantum information just linear algebra?"

One more concept behind quantum computing merits an introduction. Chapter 1 introduced the probability distribution and the Shannon entropy thereof. The probability distribution has a quantum cousin called a *quantum state*. We've seen quantum states, although I haven't called them by their name. An electron spin, for example, can be in the quantum state of pointing upward. An atom can be in a quantum state that's a superposition of two locations.

In what sense is the quantum state the probability distribution's cousin? Suppose that Baxter plans to measure whether a qubit points upward. His detector will have some probability of reporting "yes" and some probability of reporting "no." Knowing the qubit's state, Audrey can calculate those probabilities. So, a quantum state encodes a probability distribution.

But it encodes more. Baxter can measure, instead, whether the qubit points leftward. Again, his detector will have some probability of reporting "yes" and some probability of reporting "no." Audrey can calculate these probabilities from the quantum state, as well. So, the state encodes another probability distribution. But Baxter can measure along any of infinitely many axes. Consequently, a quantum state encodes infinitely many probability distributions.

One distribution stands out from the others, though. Say that the qubit points leftward. It turns out to be in a superposition of pointing upward and pointing downward, as well as in a superposition of pointing forward and pointing backward. There's one axis—the left-right axis—of whose extremes the state isn't a superposition. That axis stands out.

Suppose that Baxter measures the non-superposition axis. The possible outcomes have probabilities whose Shannon entropy Audrey can calculate. That entropy is called the quantum state's *von Neumann entropy.* We met John von Neumann in chapter 1, where he advised Shannon to call his (Shannon's) uncertainty function an entropy. Von Neumann has his own entropy, the quantum analog of Shannon's.

Why focus on the non-superposition axis? Because doing so proves useful. The von Neumann entropy tracks the best efficiency with which we can process quantum information in certain ways—for example, compress quantum data. We encountered classical data compression in chapter 1. There, Audrey recorded a string of letters, each representing what she found her brother doing on a Wednesday night, when he was charged with guarding their headquarters. She aimed to compress the string into the fewest possible bits. The least number of bits she needed per letter was a Shannon entropy.

Suppose that Audrey accumulates not a string of bits, but a string of quantum states. She aims to compress the string into the fewest possible qubits. The minimum number of qubits needed per state is the state's von Neumann entropy. So, the von Neumann entropy echoes the Shannon entropy, as the Shannon entropy echoes the thermodynamic entropy, as we'll see in the next chapter.

Earlier in this chapter, I presented the jumble *HIGAFRNT* of letters. Have you unscrambled the word? It's the name of a coin that Audrey might find lying on the street outside her cabal's headquarters: *farthing.*

CHAPTER 4

THERMODYNAMICS

"MAY I DRIVE?"

Baxter asked without intending to; without thinking; without noticing anything except the thrumming, the smoke, the scent of oil, and the sense of purpose in the locomotive.

"Nae, lad." The driver spoke in a Cumbrian accent, barely parting his lips beneath his ruddy mustache.

Baxter slumped down in his seat and picked at a thread in his jacket. But, two minutes later, he was up and devouring the controls with his eyes again.

"Please, sir, may I try driving for a bit?"

"Nae, lad." The response came as softly as before.

"Just a moment?"

"Nae, lad."

"Just—"

"Baxter!" Audrey, having staggered up the carriage, snatched the back of her brother's jacket. "Stop harassing that poor fellow! How shall he concentrate on his work with you flitting about him like a mayfly? How shall—oh!" She took in the knobs, the wheels, the ratchets, the buttons. She remained silent for a moment, absorbed by the machinery, until the question rose unbidden: "Please, sir, may I have a go?"

IMAGINE STROLLING AROUND London during the early 1870s. You wear a suit or a dress that your great-grandparents couldn't have afforded. The outfit is made of cloth woven on looms powered by steam or water. Factories have lowered the costs of everyday goods.

You wrinkle your nose upon passing a factory. It belches smoke, which mixes with the fog that blankets London. No one will coin the term *smog* for another few years, but you'll grasp the name as soon as you hear it.

Other pedestrians brush past you, as London has grown denser, as well as wider. Country folk have been flocking to cities in search of jobs. Overcrowding has led to epidemics, including typhus, cholera, and scarlet fever. Fortunately, the epidemics have spurred the government to develop a sewage system.

You spy a skinny young man leaning against a doorway, reading the final novel written by Charles Dickens. Steam-powered printing presses have given the public access to more texts, raising literacy. That young man might become an engineer who climbs above the social station he was born into. He might have suffered, similarly to Dickens, an adolescence of labor and destitution. Dickens captured octaves of human experiences because, unprotected by child-labor laws, he lived them.

Thinking of Dickens, you remember the last book you borrowed from a circulating library: *The Innocents Abroad*, by that American humorist Mark Twain. Twain sailed to Europe and the Middle East on a steam-powered ship. You observed Damascus through his eyes not long after he did, without suffering his bruises from falling off a donkey.

The Industrial Revolution transformed society economically, socially, environmentally, sartorially, gastronomically, religiously, governmentally, hygienically, literarily, domestically, internationally, dreadfully, wonderfully, and stick-your-favorite-adjective-here-it'll-probably-fit-ly. The steam engine largely drove these changes.

The steam engine prompted the development of thermodynamics, the study of energy. We'll meet thermodynamics in this chapter, starting with the field's history. We'll then encounter types of energy and see how they transform into each other in engines. The engines described best by traditional thermodynamics stay in equilibrium—quietude in which changes occur slowly. Modern thermodynamics has progressed beyond equilibrium. All of thermodynamics rests on four laws, the second of which concerns liver—that is, entropy. As those laws developed, so did atomism, the notion that materials consist of particles too small to see. Atomism and thermodynamics provide foundational insights about the world, but thermodynamics also involves an engineering mindset, like information theory.

·{ Different Kind of Steamy Affair }·

Classical texts whisper about steam engines. The Roman engineer Vitruvius described a steam-powered device during the first century BCE. The Greek inventor Hero of Alexandria provided more detail a century later. Hero's device is called an *aeolipile*, after Aeolus, the Greek god of wind.

To visualize an aeolipile, visualize a turkey roasting above a fire (figure 4.1). Mark Twain smoked a pipe that made a right angle; stick such a pipe into the turkey's right-hand side, as though the bird wanted to smoke through its breast. Stick another pipe upside-down into the turkey's left-hand side. Place a cauldron of water above the fire. Transform the spits into thin tubes that extend outward from the turkey, bend downward, and terminate on the cauldron. You can replace the turkey with a hollow metal ball in your imagination, though I prefer to keep the turkey in mine.

The fire heats the water, which boils. Steam rises through the tubes, fills the ball (the turkey), and exits through the pipes. Steam rising through the right-hand pipe pushes the pipe downward, spinning the ball (the turkey). Steam expelled downward through the left-hand pipe exacerbates the spinning. Drizzle with gravy, and serve with mashed potatoes.

More than a millennium passed before steam engines impacted industry. When they did, however, they helped establish a new

FIGURE 4.1

meaning for the word *industry*. The English engineer Thomas Savery patented a steam-driven pump in 1698. He wanted to help miners empty their mines of water. His device broke metaphorical ground, but it also broke apart often, as the metal couldn't survive the pressure exerted by the steam. The invention also wasted a substantial amount of energy.

Thomas Newcomen improved upon Savery's engine in 1712. Newcomen was a tradesman who worked with iron. He entered into business with Savery, to use Savery's patent. The men failed to name their business "Thomas & Thomas," to the disappointment of future schoolchildren.

Enter James Watt, a Scotsman who fabricated scientific instruments. Watt figured out, in 1765, how to prevent the engine from wasting so much energy. He took out a patent, then went into business with manufacturer Matthew Boulton. Boulton lived in a city whose name I heard often while working as a research assistant in northern England after college. I visited other cities by train, switching lines where a voice recording announced the stop "*Beh*-ming-um *Neeew* Street," to the delight of an American learning the subtleties of British accents. Boulton lived in Birmingham, to which Watt moved.

Engineers invented and enhanced the steam engine. Markets, factory workers, governments, and consumers transformed the engine from a contraption into a zeitgeist. Early motor vehicles hold an allure captured by Kenneth Grahame in his 1908 children's book *The Wind in the Willows*. Three animal friends—Toad, Rat, and Mole—are riding a cart through the countryside one summer afternoon. Another traveler upends their day:

> In an instant (as it seemed) the peaceful scene was changed, and with a blast of wind and a whirl of sound that made them jump for the nearest ditch. It was on them! The "Poop-poop" rang with a brazen shout in their ears, they had a moment's glimpse of an interior of glittering plate-glass and rich morocco, and the magnificent motor-car, immense, breath-snatching, passionate, with its pilot tense and hugging his wheel, possessed all earth and air for the fraction of a second, flung an enveloping cloud of dust that blinded and enwrapped them utterly, and then dwindled to a speck in the far distance, changed back into a droning bee once more.[1]

The early motorcar appeals to one's sense of adventure, excitement at novelty, and nostalgia.* All three sensations gripped me as a child, on the Disney World attraction *Mr. Toad's Wild Ride*. Why that ride ranked as my favorite, I can't explain; perhaps the answer is fate. Entranced by a motorcar ride as a child, then growing up to study thermodynamics, I recapitulated the 1800s: after the engine

* A steam engine might not have powered the car witnessed by Toad and friends: In a steam-powered car, fuel combusts outside the engine. Fuel combusts inside the engine in today's cars and in models sold before Grahame published his book. The car in the book more likely runs on internal combustion than on steam. Still, the book captures the early motorcar's spirit and seductiveness.

debuted, scientists couldn't resist studying it. Hence the birth of thermodynamics.

Working like a Dog

Thermodynamics is the physics of energy—the forms that energy can assume and how energy can transmute between those forms. A turkey roasted on a spit like an aeolipile contains *chemical energy*, stored in bonds between atoms. Imagine that Baxter partakes of such a turkey. He can use the energy he's gained to lift a Wedgwood teapot onto a shelf. The teapot will acquire gravitational potential energy, which we encountered in chapter 2. Gravitational potential energy comes from defying the Earth's gravitational pull. Similarly, resisting a nucleus's positive charge gives a negatively charged electron *electrical energy*. If the teapot falls off the shelf, it'll acquire *kinetic energy*, or energy of motion. Energy exists in many other forms, but we'll focus on two for now.

One hunk of matter can transmit energy to another hunk in two forms: heat and work. The effect of *heat* on particles resembles the effect of milkshakes on four-year-olds: heat causes particles to jiggle around randomly and aimlessly. *Work* is coordinated energy that can be harnessed directly to achieve a goal.

When lifting the teapot, for example, Baxter draws on food-sourced chemical energy to perform work against gravity. After the teapot hits the floor, the teapot's kinetic energy is channeled into breaking bonds that kept the teapot whole, into vibrating particles in the floor, into raising the nearby air's temperature, and into creating sound. The smash accomplishes nothing useful: energy that used to be concentrated in the teapot is dispersed across the floor and the air—across many particles—from which no one could re-collect the energy to lift another teapot. The energy has *dissipated*.

Understanding heat and work, we can better understand steam engines. A steam engine is a type of *heat engine*, which turns a little heat into a little work while dissipating more heat. A heat engine can take the form of a gas in a cylinder.* In many cases, the engine interacts with three other systems. One is the system on which the engine performs work—the gears in a car, a battery charged by the engine, a Wedgwood teapot lifted by the engine, etc.

The other two systems, thermodynamicists call *heat baths*. We call them also *heat reservoirs*, but I prefer *baths*, which offers an excuse to draw shower heads and loofas in the notebook of my mind. A heat bath is a collection of many particles—too many particles for anyone to measure them all. The bath's large-scale properties, such as its temperature, barely change. Why? The bath is so large, the energy it can gain from most systems acts as a drop in the bucket. An example heat bath includes the air in Audrey's library. Audrey, like any living organism, radiates heat into the air. But the air contains so many particles, it absorbs her heat without changing temperature much.† The quintessential heat engine interacts with a hot bath and a colder bath.

Having introduced the cast, we can begin the play. The heat engine performs work by undergoing a cycle, often in four acts. We call the process a *cycle* because the engine returns to its initial state. We'll focus on the Carnot cycle (figure 4.2), named after the nineteenth-century French engineer Nicolas Léonard Sadi Carnot.

The curtain rises on the engine—the gas—compressed into a fraction of its cylinder. A piston, clamped to the cylinder's walls, imprisons the gas at the bottom of the cylinder. Through the cylinder's walls, the gas exchanges heat with the nearby hot bath. This

* As can many other physics concepts. Physicists adore models that encapsulate a concept's essence without the distraction of bells and whistles. The gas in a cylinder, or box, encapsulates many concepts. Despite its humbleness, it'll star in many more examples in this book.

† Wouldn't the temperature change if Audrey lit a fire in the fireplace? The fire would serve as another heat bath. Two baths form rivals worthy of each other, unlike a bath and a much smaller human. The air will heat up; and the fireplace will cool, unless stoked.

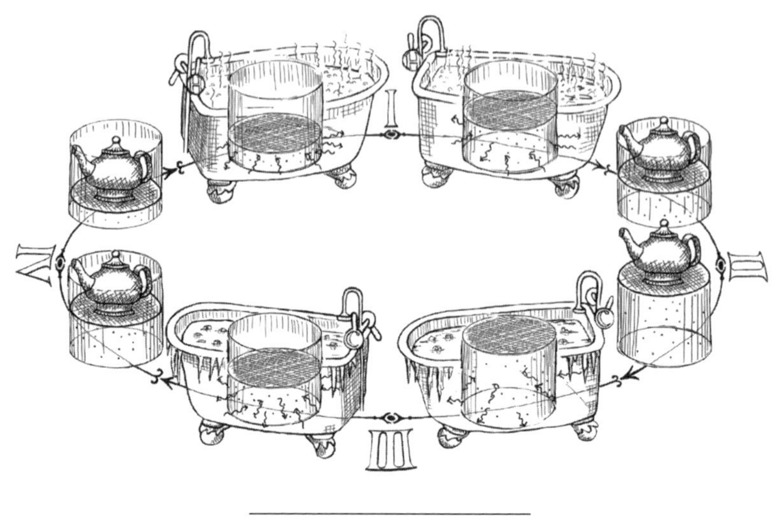

FIGURE 4.2

exchange endows the gas with the bath's temperature, which the gas will retain through act 1.

During act 1, the piston is unclamped so that it can slide. The gas, imbibing heat from the bath as a toddler imbibes milkshakes from a diner, can't stand to stay confined. The gas pushes the piston upward, expanding across most of the cylinder.

The curtain falls. A black-shirted technician carts the hot bath off the stage, while another techie places a Wedgwood teapot atop the piston. The curtain rises, and the gas expands further. Expanding—pushing the teapot upward—requires the engine to perform work against gravity. The hot bath no longer replenishes that energy, so the gas's temperature falls.

Act 2 ends. Exit teapot at its final height—carrying newly acquired gravitational potential energy. The teapot donates some of that energy toward some worthy cause, offstage. For example, the teapot may be dropped a short distance onto a ball of dough, to help flatten the dough into a biscuit.

The curtain rises on act 3 with the cylinder immersed in the cold bath. Heat leaks into the bath from the gas, whose particles slow down. They cease to beat against the piston, which begins to compress the gas into the bottom of the cylinder. The curtain descends, and the techies cart away the cold bath.

The gas's final dialogue takes place during act 4. The Wedgwood teapot has returned to the top of the piston. The teapot's weight forces the piston farther downward, performing work on the gas. The work heats up the gas, which returns to the hot bath's temperature. Once the gas returns to its initial condition, the cycle closes.

What have we gained from the engine cycle? The gas performs more work on the teapot, during act 2, than the gas receives from the teapot during act 4. Giving more than it gets, the engine *extracts work* from the baths; the audience applauds.

This work extraction comes at a cost: The engine absorbed lots of energy from the hot bath. Only a fraction of that energy turned into useful work. The engine dissipates the rest of the energy into the cold bath. Imagine running the engine cycle again and again and again. After enough runs, the cold bath will receive enough heat that its temperature will rise substantially; the cold bath will grow less cold. It will be less able to facilitate work extraction in the future. Therefore, extracting work comes at the cost of energy dissipation, which hampers future work extraction.

Work extraction is a thermodynamic task just as data compression is an information-processing task. The Shannon entropy measures the efficiency with which we can compress classical data, and the von Neumann entropy measures the efficiency with which we can compress quantum data. Thermodynamic tasks, too, have efficiencies. The heat engine's efficiency depends on two quantities: the work performed by the engine and the heat absorbed from the hot bath. Divide the first by the second, and you've defined the bang you get for your buck.

The heat engine's efficiency can't exceed a certain bound, according to Carnot, the French engineer. He proved a limit on the efficiency of any engine that exchanges heat with exactly two baths. We call that limit the *Carnot efficiency*, you might not be surprised to hear. The hotter the hot bath, and the colder the cold bath, the greater the Carnot efficiency. Given an infinite-temperature bath and a zero-temperature bath, you can extract work with an efficiency of one. The smaller the gap between the temperatures, the closer the best efficiency lies to zero. An engine running on the Carnot cycle operates at the Carnot efficiency, as one might hope.

You'd be twiddling your thumbs before the cycle closed, though. The Carnot engine operates infinitely slowly; the play that I described takes forever. Speeding up dissipates more energy than necessary, lowering the efficiency below its greatest possible value. So, no real engine is a Carnot engine, and no real engine operates at the Carnot efficiency. Then why did Carnot dream up his engine? To pinpoint a fundamental limitation on what an engine can achieve when pushed to an extreme. Thermodynamics encompasses not only practicality, but also ideality—not only the machines that power factories but also the limits on what's possible. Although thermodynamics sprouted from engineering, it's rooted in physics and chemistry.

In Praise of the Doldrums

Another thermodynamic ideality resembles the adolescence of a teenager who dreams of adventure but can't escape their small town—a situation in which nothing interesting happens, prima facie. We call this thermodynamic ideality *equilibrium*, and every system in equilibrium meets two conditions. First, the system's large-scale properties—energy, volume, temperature, etc.—remain approximately constant. They might fluctuate, but not much. Second, no net flows enter or leave the system. For example, imagine a blueberry scone on a countertop in the Stoqhardt kitchen. The scone has cooled after exiting the oven; now, as much heat leaves the scone as enters it, on average. The scone, we say, is at equilibrium with the kitchen air. Furthermore, because the scone shares the air's temperature, we say that the scone is at *thermal equilibrium* with the kitchen.

In equilibrium, nothing large-scale changes; monotony reigns, to the naked eye. But, if we zoom in, we'll find plenty to attract our attention. Particles zoom hither and yon, collide, and change direction. (Classical physics describes the particles, as they have a high temperature and low density.) At each instant, the collection of particles is in some *microstate*, defined by a list of the particles' positions and momenta (and, depending on what the particles consist of, their angular momenta, their vibrations, etc.). The microstate changes quickly, due to the zooming hither and yon and the collisions. The collection of particles is also in some *macrostate*, which consists of the collection's large-scale properties: the collection's energy, the collection's volume, the number of particles in the collection, etc.

Suppose that you know that a collection of particles is in some macrostate—that the collection has some energy, volume, number of particles, and so on. In that macrostate, many microstates are possible. An air molecule in the Stoqhardts' library can linger above a dictionary, or leftward of the dictionary, or farther leftward, while the air, as a whole, has a fixed energy, a fixed volume, and so on. The air has some probability of being in *this* microstate and some probability of *that* microstate. All the microstates' probabilities equal each other, in equilibrium. These probabilities form a probability distribution, of the sort we saw in chapter 1. There, we associated probability distributions with entropies. Shall we associate our probability distribution over microstates with a liver—I mean, an entropy? Indeed, we shall, in a few pages.

⁘{ Escape from the Doldrums }⁘

Not all systems are in equilibrium, as illustrated by a bumper sticker in my high school chemistry classroom. "Old chemists don't die," the sticker said. "They reach equilibrium." We, as organisms, live far from equilibrium. Our temperatures don't match our environments', unless by coincidence. At meals, we consume energy without releasing much. We compensate afterward by radiating more heat than we soak up.

We remain far from equilibrium by interacting with other systems—such as with the Yorkshire pudding we eat for dinner. A system is doomed to equilibrate if it interacts with nothing else. We call such a system *closed and isolated*. Systems that interact with others are *open*.

Thermodynamics used to focus on equilibrium, but the field has broadened to include systems such as us. Ilya Prigogine, a Russian-turned-German-turned-Bruxellois-turned-American chemist, exemplified the trend. He won a Nobel Prize in 1977 for studying what he called *nonequilibrium thermodynamics*. I sometimes describe my research as nonequilibrium thermodynamics. The occasional colleague, who studied mostly Victorian thermodynamics, will blink and ask whether I've used an oxymoron. But the term *nonequilibrium thermodynamics* appears in Prigogine's Nobel Prize lecture.[2] Prigogine defied convention not only in his terminology but also by pursuing history, archaeology, and piano.[3] To gain insight into open systems, he demonstrates, one should keep an open mind.

Before Prigogine arrived on the scene, his predecessors codified four laws of thermodynamics.* We'll discuss three of these laws in this section and the final law in the next section. Naturally, the laws

* A scientific law is a statement believed to be true because it's withstood many experimental tests. Some future experiment might break the law, in which case we'll search for a more accurate law.

begin at number zero. The zeroth law took shape in 1939, crystallized by British scientists Ralph Fowler and Edward Guggenheim.[4] The first, second, and third laws had gained acceptance decades earlier. Fowler, however, decided that the new law deserved to head the pack.[5] I numbered this book's prologue chapter 0 to honor Fowler's gumption.

The zeroth law establishes the concept of a thermometer. Suppose that Baxter has a spoon that's at thermal equilibrium with an almond pudding being eaten by Audrey and with a curry being eaten by Caspian. Audrey's almond pudding, says the zeroth law, is in thermal equilibrium with Caspian's curry. Systems at thermal equilibrium with each other have the same temperature, we've seen. Suppose that Caspian knows his curry's temperature. Baxter's system acts as a thermometer, with which the friends diagnose the temperature of Audrey's almond pudding.

The first law of thermodynamics formalizes the conservation of energy: Every closed, isolated system's energy remains constant. Any change in a system's energy comes from heat or work or both.

The second law of thermodynamics deserves its own chapter—no, book—no, billboard swathed in neon lights. The second law, like the actor in a one-man show, assumes many guises. In one guise, the second law states that heat can't flow from a cold system to a hotter system while the rest of the universe remains unchanged. But, you might protest, some heaters pump heat from cold systems to hotter systems. If they didn't, I would have turned into a popsicle attending college in New Hampshire. Why didn't I? Because heaters perform work to direct heat from colder systems to hotter. Because of the work spent, the rest of the universe doesn't remain unchanged.

The second law's best-known guise involves an entropy—yes, another entropy. The thermodynamic entropy predates the entropies we've encountered—the Shannon entropy in classical information theory and the von Neumann entropy in quantum information theory. Imagine a thermodynamic system—say, the steam rising from a strawberry-and-rhubarb pie in the Stoqhardts' kitchen. The steam is in one of many possible microstates. Each possible microstate has some probability of being the steam's actual microstate. Those probabilities form a distribution that has a Shannon entropy.

We can transform the Shannon entropy into a thermodynamic entropy by multiplying the Shannon entropy by another number, called *Boltzmann's constant* and represented by k_B. Ludwig Boltzmann was an Austrian physicist who cofounded thermodynamics and whose beard could have rivaled a wizard's. The numerical value of Boltzmann's constant won't matter to us; k_B is just some number hardwired into our universe, like the electron's mass. Multiply the Shannon entropy by k_B, and you have a thermodynamic entropy.*

How does the steam's entropy change? Often, the more microstates the steam can access, the greater the steam's entropy. The steam begins close to the pie, so each particle can be in one of just a few positions. The steam can access few microstates, so the entropy begins low.

Time passes, and the steam expands. Particles can reach the floor, then the ceiling, then the nearest wall. Every steam particle occupies one of many positions. The number of microstates mushrooms, and the entropy toadstools.

Now, the particles reach the far walls, bounce off, and explore the corners. The number of accessible microstates quits growing as the steam reaches equilibrium. The entropy flatlines.

The second law formalizes the behavior illustrated by the steam: every closed, isolated system's entropy grows or stays

* If the steam is out of equilibrium, some thermodynamicists disagree about whether we should call the product a thermodynamic entropy. But many thermodynamicists agree, leveraging mathematical and conceptual arguments.

constant; the entropy can't decrease. The second law explains two phenomena that we've encountered and one phenomenon that I've alluded to. The first phenomenon that we've encountered is the Carnot efficiency; Sadi Carnot upper-bounded the efficiencies of all engines that contact exactly two heat baths. His "no engine can exceed this efficiency" statement derives from the second law's "no closed, isolated system's entropy can decrease."

The second phenomenon that we've encountered is the Wedgwood teapot that falls off a shelf and shatters on the floor. The teapot's kinetic energy dissipated as heat, noise, and floor vibrations. We can now understand why. The teapot-and-air-and-floor system can access more microstates if the energy dissipates. Any given quantum of energy, by dissipating, can drive *this* air particle or *that* air particle, or vibrate *this* chunk of the floor or break *that* chemical bond in the teapot. The quantum of energy enjoys oodles of options, so the system can access oodles of microstates, so the entropy grows. The energy dissipates in accordance with the second law of thermodynamics.

The energy needn't dissipate according to other laws of physics. For example, consider Newton's laws of classical mechanics or the laws of quantum theory. These laws contain equations that govern how a system changes as time progresses. If time flowed backward, the equations would look identical. So, physical laws set the future and the past on equal footing.

According to Newton's laws, after the teapot smashes, its energy can bounce back from the air and the floor into the teapot. The chemical bonds in the china can repair themselves; the teapot can fly back onto its shelf. A smashed-teapot microstate can evolve into a reassembled-teapot microstate. But loads more microstates are consistent with the teapot's staying smashed than with the teapot's reassembling. In equilibrium, all the microstates are equally probable next steps. So the most probable macrostate—"teapot stays smashed" or "teapot reassembles"—corresponds to the most microstates. The "teapot stays smashed" macrostate corresponds to the most microstates, we argued above. Therefore, Baxter's mother will probably never take her tea in her favorite china again.

The teapot story reveals why we should give a fig about the second law. Not only does the second law raise our heating bills and curtail our cars, but the second law also decrees that time flows in only one direction. This is the phenomenon that I've only alluded to so far that the second law of thermodynamics explains: time can't run backward. We can't reverse into yesterday as the Stoqhardts' driver backs a carriage into a shed. Teapots smash; leaves transform from green to brown; and, as J. Alfred Prufrock, the speaker of the poem from chapter 2, laments:

> I grow old . . . I grow old . . .
> I shall wear the bottoms of my trousers rolled.

This decree has hit home more than I expected: I keep encountering graveyards on trips I've taken for the sake of thermodynamics. The trend began early in graduate school, at a conference at the University of Cambridge. Entropy starred in most of the talks. One afternoon, the conference organizers walked us participants to the Ascension Parish Burial Ground, on the city's outskirts. We wended our way among Nobel laureates' graves, searching for Sir Arthur Eddington. Eddington was the astronomer who transformed *Albert Einstein* into a household name. His 1919 observation of a solar eclipse supported Einstein's theory of general relativity. Four years earlier, Eddington had written in his book *The Nature of the Physical World*:

> The law that entropy always increases—the second law of thermodynamics—holds, I think, the supreme position among the laws of Nature. If someone points out to you that your pet theory of the universe is in disagreement with Maxwell's equations [of electrodynamics]—then so much the worse for Maxwell's equations. If it is found to be contradicted

by observations—well, these experimentalists do bungle things sometimes. But if your theory is found to be against the second law of thermodynamics, I can give you no hope; there is nothing for it but to collapse in deepest humiliation.[6]

The cemetery encounters didn't end at Cambridge. They've repeated as I've visited Banff, Canada, for information theory conferences; in Oxford University, at a quantum thermodynamics conference; and at Yale University, where I presented about quantum steampunk. Message received, universe; the second law isn't merely academic chaff.

Another quotation captures the second law's significance to this book. The quote bridges thermodynamics with quantum computation and was written by Seth Lloyd. We met Seth in chapter 3, as a cofounder of quantum computation. He also cofounded quantum thermodynamics. While echoing Eddington, Seth appeals to the modern citizen: "Nothing in life is certain except death, taxes and the second law of thermodynamics."[7]

{ Third Law's the Charm }

After the brass musical band that is the second law of thermodynamics, the third law feels like a downtempo-ambient-chill mixtape. But the third law, like its compatriots, deserves our attention. Imagine cooling a system—say, the air in the Stoqhardts' kitchen. Molecules slow down and quit ramming into each other like knights at a joust. The temperature lowers and lowers toward absolute zero. It never arrives, according to the third law; no process (of finitely many steps) can cool a system to absolute-zero temperature.

We can turn the statement around to understand it better: Imagine that we've managed to cool a system to absolute zero. Every system has a *heat capacity*, which dictates the amount of heat required to raise the system's temperature by one degree. A system's heat capacity approaches zero as the system's temperature approaches zero. So, if a system could arrive at absolute zero, an infinitesimal amount of heat—the teensiest amount imaginable—would raise the system's temperature. No system can hide from such amounts of heat. So, the system's temperature would rise; the temperature couldn't stay at absolute zero.

The concept of absolute-zero temperature merits a closer look. The founders of thermodynamics didn't know about quantum mechanics, but we do, thanks to chapter 2. Cooling a system to near absolute zero stamps out its classical behaviors. The system descends its energy ladder, which looks more like a slide—a continuous surface—near the top. But the rungs assert their discreteness at low temperatures. At absolute zero, the system stands on its lowest-energy rung.

Experimentalists wish that they could cool systems to absolute zero—for instance, to suppress quantum computers' decoherence. How low can we go? Physicists specify temperatures in Kelvins. The Kelvin is a unit of temperature, like degrees Fahrenheit or degrees Celsius. You and I have temperatures of about 310 Kelvins; outer space has a temperature of 2.7 Kelvins. MIT experimentalists have cooled molecules to 22 millionths of a Kelvin.[8] That temperature calls for a pat on the back—or the scientific version of a pat on the back, a publication in the journal *Nature*—but the third law wags a finger if we set our sights too low.

{ Only Atoms and the Void }

We've now met all the laws of thermodynamics, from zero to three. We can phrase the laws without reference to microstates, particles, or probabilities. I referred to those things when explaining the second law for convenience. Picturing particles is useful when you know that matter consists of particles. The founders of thermodynamics didn't, although for millennia thinkers had theorized about the existence of atoms.

Atomism emerged in ancient Greece, close to when Socrates questioned Athens into a tizzy. The word *atom* derives from Greek and means "cannot be cut." Matter consists of basic, uncuttable units, according to the philosophers Leucippus and Democritus. They taught that the world consists of two components: atoms and the void. Atoms' shapes and sizes determine how they can fit together, forming clusters. The clusters determine the macroscopic world's structure and properties; changes in the macroscopic world stem from atoms' random collisions.

Atomism has fallen out of and back in favor across the centuries. The philosophy transformed into science by the early 1800s, thanks largely to John Dalton. Dalton was born about an hour-and-a-half's drive northwest of Lancaster University, where I worked as a research assistant after college. Dalton practiced chemistry, which had recently sprouted from the fertilizer of alchemy. According to his theory, atoms differ between elements but are identical if they belong to the same element. Compounds, such as water, he wrote, consist of atoms clumped together in groups. Dalton's theory enabled him to predict how much mass one atom carried relative to another. Evidence supported much of his theory, which now permeates high school chemistry classes.

I met Dalton, so to speak, the year I lived in Lancaster. One spring weekend, I visited Manchester, an hour's train ride southeastward. Not knowing a soul there, I delighted in recognizing one name: a statue of Dalton sits in the town hall, as he taught and practiced science in Manchester. The location seems fitting, as Manchester industrialized before any other city. Steam powered its cotton mills during the early 1800s—steam that raised a controversy among thermodynamicists about atoms.

Some thermodynamicists embraced atomism. Examples include James Clerk Maxwell, who established the theory of electrodynamics (chapter 2), and Ludwig Boltzmann, who wore the wizard-like beard. Maxwell and Boltzmann invented my description of the steam rising from a strawberry-and-rhubarb pie: they envisioned a gas as a collection of (classical) particles that have various probabilities of being *here* and *there* and *over there*.

Not all of Maxwell's and Boltzmann's contemporaries bought the story. Thermodynamics had focused, until then, on macroscopic properties: temperature, pressure, volume, energy, and so on. The theory had triumphed despite remaining mum about matter's constituents. So why posit the existence of itsy-bitsy particles that no one can see? critics asked. Stick to what we know exists.

Evidence of atoms accumulated like dust bunnies under a bed, although we haven't room to review the evidence in this book. Physicists ended up with no choice but to embrace Maxwell and Boltzmann's theory. It grew into the field of *statistical mechanics*, the tubes and belts under the hood of thermodynamics.

How does Leucippus and Democritus's legacy fare today? We perpetuate their terminology, using the word *atoms*. What we call atoms, though, can be "cut" into smaller constituents—protons, neutrons, and electrons. The protons and neutrons can be cut into yet smaller constituents, called quarks. But quarks and electrons resemble what the Greeks called atoms—fundamental particles. I hesitate to add that these fundamental particles can't be split, for two reasons. First, electrons aren't particles in the sense of being teensy balls; they have wavelike properties, as discussed in chapter 2. Whether a wave can or can't be cut is a question that doesn't necessarily make sense. Second, experimentalists have separated an electron's charge from its spin, and observed fractions of an electron's charge, under special conditions. So, quarks and electrons aren't as fundamental as one might think, despite being fundamental particles.

Fractional electron charges fall under the purview of statistical mechanics, which has expanded since its founding by Maxwell,

Boltzmann, and others. Statistical mechanics describes not only many-particle systems that we can approximate as classical, such as steam. Other statistical-mechanical systems require quantum descriptions that Maxwell and Boltzmann didn't dream of. Quantum statistical mechanics features quantum states as classical statistical mechanics features probabilities of microstates.

Invoking probabilities, states, and particles distinguishes classical statistical mechanics from thermodynamics, according to some. I used to agree, until I began exploring information theory. Information theory is *operational*, focusing on how efficiently an agent can leverage resources to perform tasks. For instance, how many bits must you use to send a message down a wire to a friend? Information theory shares its operationalism with thermodynamics. Thermodynamicists calculate how efficiently engines can pump water out of mines, how an engine's efficiency trades off with its power, and how much heat we waste—how efficiently an agent can leverage resources to perform thermodynamic tasks. Work, heat, and efficiency—quantities that are useful or detrimental—serve as nuts and bolts in the thermodynamicist's conceptual toolbox. Granted, thermodynamicists also ask fundamental questions, such as why we age while some materials resist equilibration. But the fundamental questions intertwine with the operational ones, and each set of questions informs the other. So, I see the thermodynamicist's operational mindset as distinguishing thermodynamics from statistical mechanics. I'll therefore claim to discuss thermodynamics throughout this book, despite invoking probabilities and quantum states.

One more tool shall we wield as thermodynamicists bent on operational tasks. To grasp the tool, let's imagine the steam rising from an apple tartlet fresh from the oven. Suppose that we've captured the steam in a bottle. We cap the bottle with a piston, a thin covering that we can push downward or let slide upward. We want to compress the gas into the bottle's bottom half. To push the piston downward, we have to fight the gas particles that beat against it. How much work must we perform? Let's assume that the gas can exchange heat, through the bottle, with the rest of the room. Compression costs the least work if we push infinitely slowly. Speed up, and we'll spend work on roiling the gas, which will dissipate heat into the air.

When at thermal equilibrium with the rest of the room—when at the same temperature as the room—the gas has a property called *free energy*. We can think of the free energy roughly as the amount of work that the gas can perform. The uncompressed gas has a little free energy. The compressed gas has more free energy, like a coiled spring, being able to punch the piston to the top of the cylinder if allowed. The gap between the two free energies is the least amount of work needed to compress the gas—a measure of the resources required.

The more energy the gas has (for example, the more kinetic energy the gas has), the greater its free energy. Also, the more thermodynamic entropy the gas has, the less its free energy. Since the thermodynamic entropy determines the free energy, and since free energies determine the resources required to compress the gas, an entropy helps determine the resources required for a thermodynamic task. Entropies determined the resources required for information-processing tasks—classical and quantum data compression—in chapters 2 and 3. No wonder information theory, quantum physics, and thermodynamics have cozied up to each other. We've now surveyed all three subjects, observing parallels among them. In the next chapter, we'll mash the subjects together and start watching them interact.

CHAPTER 5

A FINE MERGER

THERMODYNAMICS, INFORMATION THEORY, AND QUANTUM PHYSICS

Other people had friendships; Audrey had a trade agreement with Lillian Quincy. Lillian kept Audrey abreast of trends in politics and philosophy, and Audrey explained the latest experiments and inventions. Today, Lillian served her guest cakes topped with lemon curd, a breath of sunshine on a rainy afternoon.

"I know you never attend the Royal Society's public lectures," Lillian said, setting her plate down, "but they are marvelous this season. Mr. Raja is lecturing, and his explanation of the quantum vacuum was exquisite. I could almost see how the vacuum contains nothing and yet has energy nonetheless."

Audrey had gleaned that Lillian's admiration for Mr. Raja didn't stop at the quantum vacuum. Lillian had admired his innovation, analogies, and manner before; and when Audrey had explained a discovery that he'd made, Lillian's eyes lit up. Nor would Mr. Raja remain indifferent to Lillian, Audrey expected. He'd praised a drawing of hers that hung in the Stoqhardt manor; his father had come from the same neighborhood in Madras as Lillian's mother; and Mr. Raja fancied German Romanticism as much as Lillian did. Audrey bit into her lemon cake. If he found his way to the salon hosted by Lillian's mother, Audrey would soon be visiting not Miss Quincy but Mrs. Raja.

HAVE YOU EVER had two friends who haven't met but should; who share a host of qualities, interests, and values; and who seem destined to ride into the sunset together? Such are thermodynamics and quantum computation. They share an operationalist philosophy, highlighting which tasks an agent can accomplish if given limited resources. Quantum-information-processing tasks include information compression, while thermodynamic tasks include gas compression. Additionally, both theories reach across multiple disciplines. Quantum computation draws physicists, engineers, chemists, computer scientists, and mathematicians, while thermodynamics governs physics, engineering, chemistry, astronomy, and biology. Entropies star in each field, helping to determine the optimal efficiency with which an agent can perform tasks. So, thermodynamics and quantum computation merit one of those heart-shaped picture frames. Made from steel. Decorated with brass-colored decals shaped like gears.

But a marriage—a relationship that thrives and lasts—requires more than similarities. It requires each partner to enhance and enrich the other. Thermodynamics and (quantum) computation meet this standard: information can serve as a sort of thermodynamic fuel, and thermodynamic work can reset information. This chapter highlights how thermodynamics intertwines with information processing and how quantum physics transforms both. We'll start with an engine that turns useless heat into useful work, with help from information. Running the engine backward, we can pay thermodynamic work to reset information. The engine benefits from quantum phenomena including entanglement. By invoking information, the engine shows, we can resolve one of the oldest paradoxes in thermodynamics.

From Useless to Useful

Leo Szilard discovered that information can serve as a sort of thermodynamic fuel. Szilard was a Hungarian-turned-American who practiced physics during the twentieth century. He joins John von Neumann and Alfréd Rényi as a Hungarian mind who powers much of this book. In 1929, Szilard showed how to use information to convert heat into work.[1] Heat is random energy that isn't

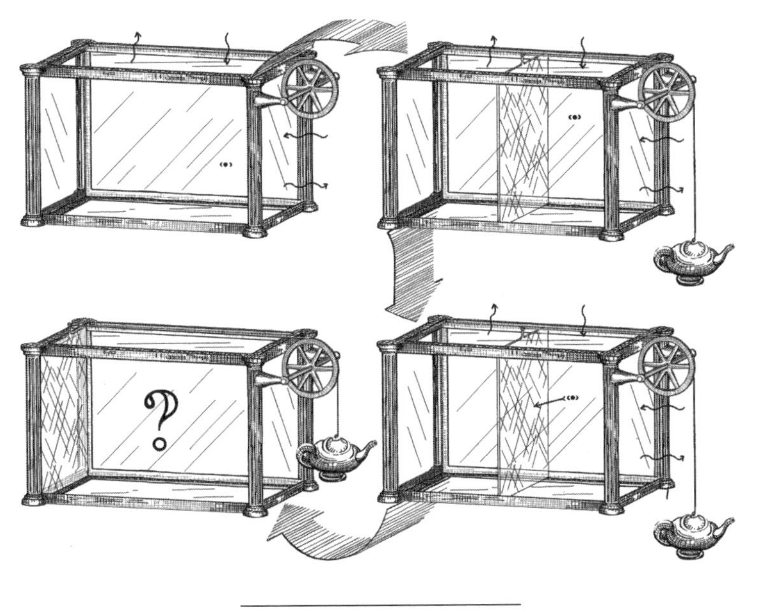

FIGURE 5.1

accomplishing anything. Work is useful, coordinated energy that can, say, turn a printing press's crank. So, information can convert useless energy into useful energy. The demonstration consisted of a thought experiment—an experiment performed in our imaginations—and an analysis. Experiments in the physical world didn't follow until the twenty-first century.

Suppose that Audrey has a gas in a box, as in figure 5.1. I warned that physicists adore gases in boxes, which can capture a problem's essence without complications. Szilard pushed this simplification to its extreme, envisioning a gas of one particle. The particle is classical, as Szilard was illustrating the relationship between classical information and thermodynamics. We'll see later how his ideas change if the particle is quantum. The box sits in a heat bath that has a temperature T. The bath exchanges heat with the gas through the box's walls.

Audrey slides a thin partition into the box's center, then measures which side of the partition the particle occupies. Never mind how Audrey measures the particle; Szilard's story is powerful because it doesn't depend on such minutiae, which depend on Audrey's personal preferences and which engineers call implementation details. I'll omit unimportant implementation details, focusing on crucial design elements, throughout this story. For concreteness, suppose that the particle occupies the box's right-hand side. Audrey has acquired one bit of information: right, rather than left.

Audrey ties a rope to the top of the partition, passes the rope over the box's right-hand side, and runs the rope through a pulley. Then, she ties a Wedgwood teapot to the dangling end of the rope.* Audrey unfixes the partition, so that it can slide leftward or rightward within the box. The gas expands, like the steam that emanates from an apple tartlet and fills a kitchen. Audrey's gas particle hits the partition, punching it rightward again and again.† Nothing punches the partition from the left, so the partition eventually reaches the box's left-hand side. Any energy lost by the gas, in punching the partition, is replenished through heat from the bath. Audrey removes the partition from the box.

Two features of this outcome draw our attention. First, the particle can now be anywhere in the box. Audrey has no idea where it is; she's lost her bit of information. Second, as the partition shifted, the teapot rose: the gas performed work against gravity. Why did

* The teapot needs to be far smaller than the Wedgwoods you can purchase at a department store, as we'll see.

† The partition must not weigh much, as it gives way to one particle. Never mind how we'd craft such a light partition; in a thought experiment, the implementation details matter less than the overall concept. Furthermore, experimentalists can realize Szilard's idea, as we'll see below.

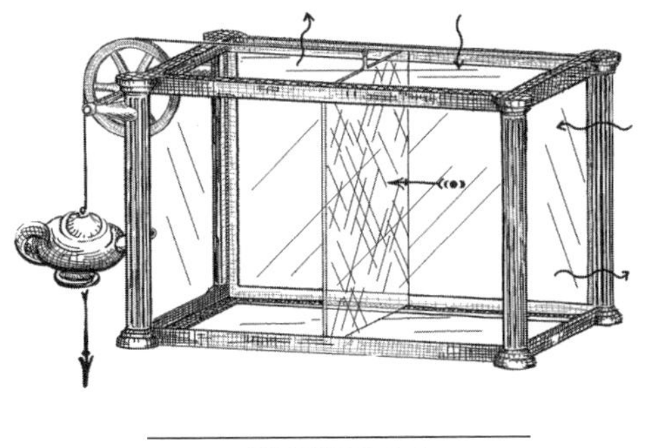

FIGURE 5.2

the teapot rise? Because Audrey ran the rope over the box's right-hand side, because she knew that the particle initially occupied the right-hand side. Suppose that Audrey hadn't acquired her bit of information about the particle's location. She'd have had to guess where the particle was, right or left. She'd have had a 50% chance of running the rope over the box's left-hand side. If she'd done so, the gas's expansion would have lowered the teapot, depriving the teapot of gravitational potential energy (figure 5.2). So, as Audrey actually ran the rope over the box's right-hand side, she leveraged information to perform work on the teapot.

Whence came the work? Not from the particle, which has the same temperature—and so the same amount of kinetic energy—as it had originally. Heat from the bath transformed into work, with help from Audrey's bit of information. She traded information for work—a computational resource for a thermodynamic resource.

How much work can the gas perform? The amount depends on two important quantities—first, the bath's temperature. The hotter the bath, the more energy it can give the teapot, so the more work Szilard's engine can perform on the teapot. Second, the work performable by the gas depends on how much the particle's Shannon entropy grows—how much Audrey's uncertainty about the particle's location grows. At the end of the experiment, the particle can be anywhere in the box; the particle has a 50% probability of occupying the box's left-hand side and a 50% probability of occupying the right-hand side. Audrey ends the experiment with the greatest possible uncertainty about the particle's position. The probabilities have a Shannon entropy of one bit, according to chapter 1. Immediately after Audrey's measurement, the particle had a 0% chance of occupying the left-hand side and a 100% chance of occupying the right-hand side. This earlier distribution had a Shannon entropy of zero, as Audrey knew the particle's position with certainty. So, as the gas expands, the entropy grows by one bit. The entropy's growth—Audrey's loss of knowledge—counterbalances the teapot's gain in gravitational potential energy. So, the amount of work performable by the gas is proportional to the entropy's growth.

I'll call this amount of work a *szilard*. If the gas is at room temperature, a szilard is an eensy-weensy amount of energy: an incandescent light bulb radiates 10^{22} times more energy per second. Suppose that Audrey wanted to lift the teapot about a yard, using a szilard of work. The teapot would have to weigh as little as 500 of the molecules that power our body's cells (adenosine triphosphate, or ATP). If you requested such a teapot at a department store, the employee behind the counter would raise an eyebrow. An artisan in a quantum-steampunk novel, though, would fill the order in a week.

Another engine lifted a Wedgwood teapot in chapter 4—a Carnot engine that contacts a hot bath and a cold bath at different times. How does Carnot's engine compare to Szilard's engine? Szilard's engine involves one bath, but we can regard Audrey's information loosely as a cold bath. When Audrey measured the particle's location, obtaining a bit of information, the particle occupied the right-hand side. Let's label the box's right-hand side as 0 and the left-hand side as 1. Audrey's bit was 0. A refrigerator can cool a qubit to the quantum analog of a 0 bit, as we saw in chapter 3. Just as a 0 qubit is cold, we can think of a 0 bit as cold.

Therefore, Szilard's engine resembles Carnot's engine, with *information* replacing the cold bath.

•{ GAME, RESET, MATCH }•

What happens if we reverse Szilard's engine? Imagine that Audrey removes the partition from the box and leaves the room, whereupon Baxter wanders in. He finds the gas spread throughout the box. The particle has a 50% chance of occupying the box's left-hand side and a 50% chance of occupying the right-hand side. The probabilities' Shannon entropy peaks, at one bit. Baxter wants to return the particle to the box's right-hand side, to reset the particle to a known position. The probabilities will change from 50% and 50%, to 0% and 100%—and to having no entropy.

We can regard the resetting as erasure. Imagine receiving a sheet of paper filled with scribbles. Consider running an eraser all across the sheet, resetting the sheet to a clean, usable state. This erasure parallels the resetting of the particle to a known position untainted by entropy. We call bit reset *Landauer erasure*, after the twentieth-century physicist Rolf Landauer who worked at IBM. Landauer realized that erasing a bit costs thermodynamic work.

Baxter resets the bit by sliding a thin partition into the box, beside the left-hand wall (figure 5.3). He pushes the partition rightward until it reaches the box's center. Baxter has compressed the gas into the box's right-hand side, so he knows that the particle occupies the box's right-hand side, rather than its left-hand side. Therefore, Baxter has gained one bit of information about the particle's position. Compressing the gas costs work, and *Landauer's principle* states how much:[2] Baxter performs at least as much work as Audrey obtained from expanding the gas, one szilard. Whereas Audrey spent information to perform work, Baxter spends work to obtain information.

Landauer's principle reveals that thermodynamics and computa-

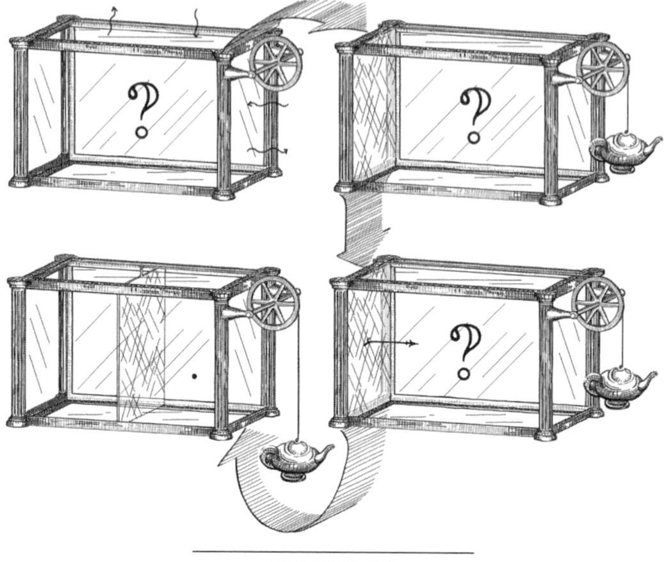

FIGURE 5.3

tion interlace like the gears in a clock. Suppose that we want to compute . . . and compute and compute and compute. Computation requires scrap paper, which we'll run out of. To continue computing, we'll have to erase used scrap paper. Erasure costs work, according to Landauer's principle. So, computation can't escape thermodynamics. Who would have thought so, a priori? Calculating your taxes seems unconnected to the science of steam engines. But the two turn out to be intertwined.

This unexpected intertwining entranced me when I first encountered Szilard's engine, in my first quantum-computation course. My professor covered Szilard's engine late in the spring, after we'd chosen our final projects. I wished that he'd explained earlier, so that I could have centered my project on the engine. Not that I knew which questions to ask or where I wanted to drive the engine; I only knew that I craved more.

Instead of a four-week project, I've centered my research

program on the spark in that lecture. Today, I can better pinpoint the source of the engine's enchantment: information is abstract, something we can't touch. Granted, Audrey can touch the pencil that encodes information about Caspian's visit, and you can grasp a book in which these words are printed. But physical systems only *encode* information; they aren't information. Despite being intangible, though, information can help raise a teapot, as physical an object as ever graced this Earth. So, an abstraction can impact the physical world. This tension also underlay my fascination with entropy in ninth-grade biology class: a funny-looking function—an ugly duckling of a mathematical object—explains why time passes, why trees sprout and fingernails grow.

Abstraction and physicality square off more famously in the mind-body problem: Minds seem immaterial because they're emergent phenomena. Yet minds influence matter, as when one adds a line to one's blueprint for a flying machine. Furthermore, minds exist only in the presence of certain matter. This tension between materiality and immateriality has fascinated philosophers, scientists, writers, and twelve-year-olds for millennia. The intersection of thermodynamics and information deserves no less limelight.

We've concluded that erasure costs work, but why should it? Why should anyone have expected our argument's conclusion? The work turns into heat that dissipates into the heat bath, raising the universe's entropy. The rise of entropy accompanies the flow of time, which enables teapots to shatter but not unshatter—processes to occur but not unoccur. Erasure is a process that can't unoccur, an *irreversible* process: you can't recover the information you've erased. Having forced the particle into the box's right-hand side, you can't infer where the particle began.* So, erasure is

* Irreversibility contrasts with *reversibility,* exhibited by operations such as the addition of 1. Add 1 to any real number, and you can infer the original number. If adding 1 yields 267, the process began with 266.

irreversible; irreversibility accompanies the flow of time; the flow of time accompanies entropy growth; and entropy grows when work is dissipated. We've established a chain linking erasure to work. So, spending work to erase makes sense.

Charles Bennett illuminated from another angle why irreversible processes should cost work. Bennett is a physicist who works at IBM, as Landauer did. He wrote in 1987 that an erasure "compressed many logical states into one, much as a piston might compress a gas."[3] By "compressed many logical states into one," he means, regardless of whether the particle began on the box's right-hand side or left-hand side—whether the particle represented a 0 bit or a 1 bit—the particle ended on the right-hand side, representing a 0 bit. The "many logical states" of 0 and 1 were "compressed" into the single logical state of 0. Logical states are represented by physical systems, such as gases in boxes. Compressing a gas costs work, according to chapter 4. Since compressing a physical system costs work, and a physical system encodes information, compressing information—erasing—should cost work.

Think Outside the (Classical Gas in a) Box

We've been discussing a classical gas in a box, but this book is called *Quantum Steampunk*, consarn it (as our nineteenth-century forbears would have exclaimed). What if the gas isn't classical? Quantum phenomena enhance information processing, and Szilard's engine processes information. Can quantum phenomena alter Szilard's engine? Yes, in many ways, three of which we'll explore.

First, inserting the partition can cost work.[4] The particle, if classical, isn't affected by the insertion. I'm assuming that the particle is small and the partition is thin, so that the partition has essentially no chance of landing atop the particle. If the particle isn't at the box's center, the particle will notice the partition's arrival as much as you'd notice a giant Pacific octopus off the Japanese coast while in Paris and avoiding news outlets.

You—and a classical gas particle—are *localized*, or confined to a finite volume. If Audrey's gas particle is quantum, it isn't necessarily localized. It's in a quantum state, which has wavelike properties that can extend across the box. In most quantum states, the particle reacts to the partition insertion, even if a measurement of the particle's location would likely report a position far from the box's center. The partition *disturbs* the particle as a measurement would. This disturbance can change the particle's energy, so inserting the partition can cost work.

Second, the particle can entangle with another system.[5] The following example riffs off of Landauer erasure rather than Szilard's engine; but Landauer erasure results from running Szilard's engine backward, so never mind the difference. Suppose that Audrey doesn't want to reset a classical particle's position, erasing a bit to 0. She wants to reset a qubit, erasing its state to the quantum analog of 0. Suppose that her qubit shares entanglement with a qubit of Baxter's. Using that entanglement, the siblings can erase Audrey's qubit while extracting work.

This result sounds like it should make Landauer roll in his grave. According to Landauer, erasure costs work rather than providing work. But the siblings can have their lemon-curd cake and eat it, too, by using the entanglement.

Say that Audrey's qubit is maximally entangled with Baxter's. According to chapter 2, the siblings can measure the qubit pair in such a way that they can predict the outcome with certainty. Four outcomes are possible, so the siblings begin with log(4) bits of information, or 2 bits. These bits can fuel two runs of a Szilard engine. Each run produces one szilard of work, according to the beginning of this chapter. According to Landauer's principle, erasing Audrey's qubit costs one szilard of work. So, the siblings retain

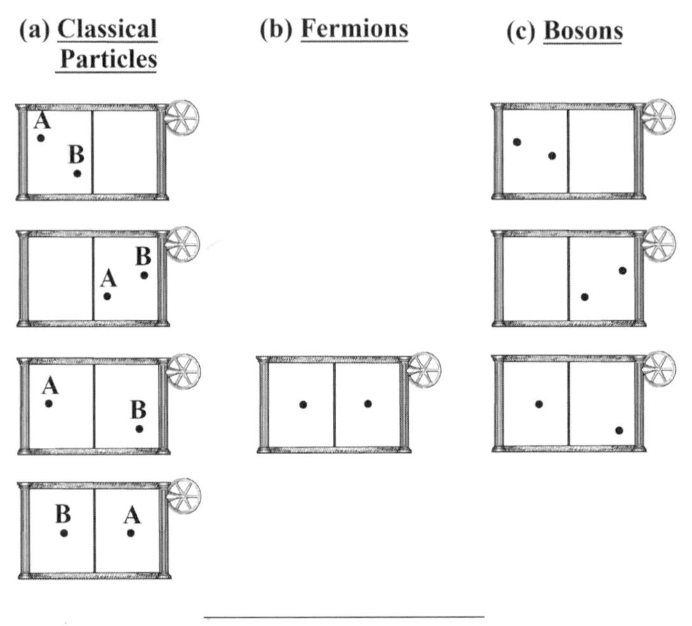

FIGURE 5.4

one szilard of work, which they can spend on lifting a teapot or running a current through a wire or whatever else they please.

Whence comes the work surplus? Audrey's and Baxter's qubits begin entangled; each qubit has quantum information about the other. At the end of the protocol, neither qubit has information about the other; the two lack entanglement. The protocol "burns" the entanglement, using quantum information (with heat from the heat bath) as a sort of thermodynamic fuel.

Third, a quantum Szilard engine that contains many gas particles can differ from a classical Szilard engine that does.[6] We can illustrate with a Szilard engine that contains two particles, one contributed by Audrey and one contributed by Baxter.

Suppose, initially, that the particles are classical. The siblings measure which side of the partition each particle occupies. One of four possible outcomes results (figure 5.4a): both particles occupy the left-hand side; both particles occupy the right-hand side; Audrey's particle occupies the left-hand side, while Baxter's occupies the right-hand side; or Audrey's occupies the right-hand side, while Baxter's occupies the left-hand side. In two of the four cases, the particles occupy different sides. The left-hand gas exerts the same pressure on the piston as the right-hand gas. The pressures balance each other, so the piston can't move; the engine can't perform work. In half the cases, the particles occupy the same side and so can perform work.

Now, suppose that the particles are quantum. Physicists have discovered two classes of quantum particles: fermions and bosons. We, and all other matter on Earth, consist of fermions. Example fermions include electrons, as well as the particles that make up protons and neutrons. Bosons transmit fundamental forces between chunks of matter. For instance, photons are bosons that transmit the electromagnetic force that attracts negatively charged electrons to positively charged protons.

Bosons tend to clump together, and fermions tend to separate. Wolfgang Pauli, a cofounder of quantum theory, identified the rule behind fermion separation. We call the rule *Pauli's exclusion principle*, which you might have learned in chemistry class. The exclusion principle explains how electrons arrange themselves in atoms. According to Pauli, no two fermions can be in the same quantum state. Say that we put two fermions in a box, slide a partition down the box's center, and measure which box halves the fermions occupy. The measurement disturbs the particles, forcing them to choose sides. Say that your measurement device detects one particle in the right-hand side. The other particle must occupy the left-hand side, so that the particles are in different quantum states.*

* I'm supposing that the fermions lack spins. If the fermions have spins, then both particles can occupy the box's right-hand side: one particle can have an upward-pointing spin, while the other particle has a downward-pointing spin. The spins will distinguish the particles' quantum states. Natural fermions have spins, but we can engineer systems that act like spinless fermions. And ignoring the spins simplifies the explanation.

Knowing how fermions behave, we can apply our knowledge to Szilard's engine. Imagine Audrey and Baxter loading a Szilard engine with two fermions, then running many trials. In each trial, the siblings measure the fermions' positions. The fermions always occupy opposite sides of the barrier, according to Pauli's principle (figure 5.4b). No pressure imbalance ever moves the barrier, so the engine never performs work.

Now, imagine the siblings running a Szilard engine with two bosons. The bosons don't always occupy opposite sides of the barrier, because the bosons don't obey Pauli's exclusion principle. But the bosons don't behave like classical particles, either. The siblings' classical particles were *distinguishable*; Audrey could always tell which particle was hers—for instance, by tracking her particle's movements. Likewise, Baxter could always identify his particle. But the bosons are *indistinguishable*; once in close quarters, they can't be told apart. So, the siblings can't identify either particle as Audrey's or Baxter's. For instance, Audrey can't distinguish a boson as hers by tracking its trajectory, because a trajectory is a sequence of positions and quantum particles lack definite positions.

Indistinguishability limits the possible outcomes of the siblings' measurement. In some trials, the siblings find both particles in the box's left-hand side; and, in other trials, in the right-hand side. Both types of trials also transpired when the siblings used classical particles. But the classical particles underwent two other types of trials, too: Audrey's particle could occupy the box's left-hand side, while Baxter's particle occupied the right-hand side, and vice versa. If the particles are bosons, neither can be labeled as Audrey's or Baxter's. The siblings can only find one boson in the box's left-hand side and one boson in the right-hand side. So, measuring the bosons yields one of only three possible outcomes, shown in figure 5.4c. In two-thirds of the possibilities, the bosons occupy the same side of the box. So, the engine performs work in two-thirds of the trials.

If a crazed engineer ever threatens to vandalize your workshop unless you extract more work than he does with a two-particle Szilard engine, request a bosonic engine. It'll perform work in two-thirds of your trials (on average), while a classical engine can perform work in half the trials (on average), and a fermionic engine never performs work. So, the bosonic engine outperforms the classical engine, which outperforms the fermionic engine. Quantum particles can beat classical particles at a thermodynamic task, just as quantum computers can beat classical computers at certain computations.

The Demon's in the Details

A quantum advantage is all well and good. But even the classical Szilard engine—or, more precisely, its reverse, the classical Landauer eraser—resolved one of the longest-standing paradoxes in thermodynamics. The paradox explains why demons feature in advertisements for thermodynamics conferences and as mascots for thermodynamics research groups. Some of the demons are red; some have horns or pointy tails; most hold tridents; and one stands in front of a chalkboard, brandishing an eraser. Not because thermodynamicists channel the medieval German sorcerer Faust, said to have conjured a demon after mastering all human knowledge. As a crown symbolizes a monarchy, a demon symbolizes James Clerk Maxwell's challenge to the second law of thermodynamics.

We've encountered Maxwell as a founder of electrodynamics and as a thermodynamicist who embraced atomism. His cap contained many feathers. One consists of a paradox formulated in 1867. Maxwell envisioned a setting familiar to us: a many-particle classical gas in a box. The gas has equilibrated; particles whiz all around the box, at various speeds. The box isolates the gas from the rest of the world.

A thin partition divides the box in half, and particles can sneak

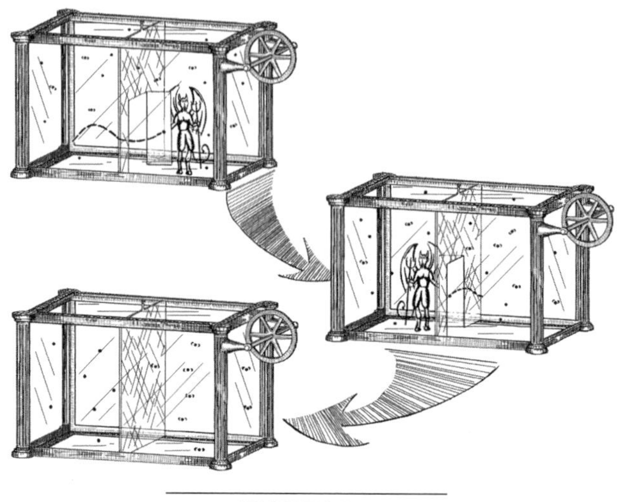

FIGURE 5.5

through a trapdoor in the partition. A "finite being," as Maxwell called it, controls the trapdoor (figure 5.5). Colleagues renamed the being a demon; hence the thermodynamicist's mascot.

The demon watches the particles whizzing around the box. If a particle approaches the partition from the left at high speed, he lets the particle through the trapdoor. He opens the door also for slow particles that approach from the right. After a while, the right-hand side contains only quick particles. The quicker a gas's particles, the greater the gas's kinetic energy, and the higher the gas's temperature; so the box's right-hand side contains a hot gas. The left-hand side contains only slow particles, which form a cold gas.

The demon has taken a mixed-up system and unmixed it. He might as well take a bowl of scone batter and remove the cream that the cook stirred in. The demon has reduced the gas's entropy. Gadzooks! But the tale darkens further.

A Carnot engine performs work, we saw in chapter 4, given a hot bath and a cold bath. The demon has created such baths, so his box can drive an engine and charge a battery. By the time the engine completes its cycle, the gas has returned to its initial state: particles whiz around randomly, at various speeds. The demon can again separate hot particles from cold, then extract work again, and repeat these two steps as many times as his black heart pleases. The demon can charge infinitely many batteries, run every power plant out of business, and lift all the teapots in the world, without paying any cost. The gas always returns to its initial state, without losing any value. The demon will run a *perpetuum mobile*, or perpetual-motion machine. Those can't exist, according to the second law of thermodynamics.

Welcome to the game that I call *Why Doesn't the Second Law of Thermodynamics Break?* As a television show, the game would collapse in flames; but it can spice up a lunch-table conversation if played with friends. One person dreams up a *perpetuum mobile*, and everyone else figures out why it can't exist. Maxwell, having initiated round one, is eating a peanut-butter sandwich as we turn to poking holes in his machine.

The second law of thermodynamics states that every closed, isolated system's entropy increases or remains constant. The gas can't interact with the outside world, so it seems closed and isolated. Or does it? The gas interacts with the demon. Does the demon's entropy grow, offsetting the decrease in the gas's entropy? It needn't, alas. We can replace the demon with an automatic mechanism that accomplishes the same task. The mechanism can avoid radiating the heat that an organism would omit. The mechanism can also avoid dissipating energy as friction or sound, if engineered perfectly.

The demon measures particles' speeds. Do the measurements increase the system's energy? Szilard postulated so in 1929, as quantum theory was crystallizing.[7] IBM researcher Charles Bennett proved otherwise decades later; he detailed a means of measuring particles without dissipating energy. He also resolved Maxwell's paradox, drawing on Landauer's principle, in 1982.[8]

Bennett argued as follows. The gas and the demon, together,

form a closed, isolated system. The demon contains a memory, which changes every time the demon measures a particle's speed: the demon must remember the speed long enough to open the trapdoor or to block the particle. The demon operates a *perpetuum mobile* only if his memory returns to its initial state after each cycle, without dissipating much energy.

We omitted a step from our earlier description of the cycle: the demon must erase his memory at the end. According to Landauer, erasure costs work. The erasure's work cost nixes the work performed by the Carnot engine. So, the demon nets no work from the supposed *perpetuum mobile*, on balance. An information-theoretic task—erasure—resolves a thermodynamic paradox.

Another way to phrase the resolution is: erasure dissipates energy. That dissipation raises the gas-and-demon system's entropy at least as much as the demon lowers the gas's entropy. So, the whole system's entropy doesn't decrease, and the second law reigns.

At the lunch table of my mind, Maxwell finishes his peanut-butter sandwich and zips up his lunchbox. It's your turn to invent a *perpetuum mobile*.

We can now understand why one cartoon demon brandishes an eraser: Landauer's principle of erasure resolves Maxwell's demon paradox. At least, many physicists believe that Landauer's principle has. Detractors remain, but the proposal has gained widespread acceptance. Experimentalists have checked Landauer's bound by manipulating single molecules and nanoscale magnets; and theorists have added bells and whistles to Maxwell's story.

At the beginning of the chapter, Audrey traded scientific expertise for political and philosophical news from Lillian Quincy. Likewise, information theorists can trade with thermodynamicists. Information can transform useless heat into useful work; one can pay work to obtain information; and information erasure resolves a thermodynamic paradox. We'll knot the threads that bind information theory and thermodynamics, as well as quantum physics, in the next chapter.

CHAPTER 6

THE PHYSICS OF YESTERDAY'S TOMORROW

THE LANDSCAPE OF QUANTUM STEAMPUNK

"Move over, Baxter. Audrey hasn't had a chance to look."

Caspian frowned at Baxter over the parchment map spread across the oak dining-room table. Grumbling, Baxter retreated to where a teacup was pinning down a curled-up corner. A silver bell straddled another corner; a copy of Andreas Vesalius's book of anatomy, the third; and an ancient Nubian statuette, the fourth.

The map had yellowed, the ink had faded, and the parchment smelled like a badger in need of a bath. But even a glance at the contents would drive the odor from one's mind: walled cities, ragged coastlines, and sinuous paths filled the map. Stepping forward to where Baxter had been standing, Audrey noticed icy peaks, golden deserts, and a cherubic little boy whose breath formed the west wind. She bent closer, tucking a stray hair behind her ear, and ran a finger over the blue whorls that marked the Sea of Elestrav.

"Should one of the edges not say, 'Beyond here, there be dragons'?" Audrey asked. Looking up, she saw Caspian smiling at her.

"The cartographer had no need to write the words out," he said. "There be dragons all over the map."

SZILARD'S ENGINE and Landauer erasure demonstrated that not only computation, but also quantum computation, interlaces with thermodynamics: entanglement can turn a work cost into a work gain, and bosons can boost the average amount of work performed by an engine. So, quantum resources can benefit a thermodynamicist. Quantum information theory helps us see how, providing the framework through which we understand entanglement. In turn, thermodynamics helps the quantum information scientist, who wants to learn how quantum physics differs from classical: entanglement and bosons differ from their classical counterparts by enabling one to extract more work.

Quantum steampunk encompasses this intersection of quantum physics, information theory, and thermodynamics. The field captures how quantum systems facilitate thermodynamic tasks, how this facilitation can distinguish quantum systems from classical, and how we can use quantum physics and information theory to update nineteenth-century thermodynamics. The update extends to small classical systems and to far-from-equilibrium systems. Although not quantum, such systems contrast with the large, many-particle, largely at-equilibrium systems studied by Victorian thermodynamicists.

This reimagining of thermodynamics has been underway for years. My colleagues, our academic forbears, and I have whittled, soldered, and sharpened a mathematical, conceptual, and experimental toolkit. The toolkit isn't complete, but it weighs enough to be dangerous. Some of us strap this toolkit to our backs, set out across the wilderness that separates scientific fields, and exchange ideas with other disciplines. We collaborate with chemists, condensed-matter physicists, particle physicists, biophysicists, and more. We leverage quantum-thermodynamic tools to help answer their questions; we borrow their tools for our field; we uncover questions in their fields; and their questions inspire projects of ours. This exchange between quantum-information thermodynamics and the rest of science, too, qualifies as quantum steampunk.

Other researchers have called my field *quantum thermodynamics*, and I sometimes call it *quantum-information thermodynamics*. While pursuing my PhD, though, I realized that the field has a steampunk aesthetic. The scientists of yesteryear studied and valued aesthetics, and we thermodynamicists look up to those scientists. So, I dubbed the field *quantum steampunk*, a term that today's students have taken up.

In this chapter, we'll tour the history of quantum steampunk. Then, we'll see how even basic thermodynamic concepts—work and heat—need rethinking in quantum contexts. Finally, we'll gaze upon the landscape of quantum steampunk—the collection of subfields and results that we'll explore over the next several chapters. During that exploration, we'll exercise our rethinking of quantum work and heat.

In the Beginning

Detailing the field's history would require another book, but I'll sketch the development of quantum thermodynamics here. The field's roots stretch back to the 1930s, the childhood of quantum theory. Physicists hoped to use quantum uncertainty—the impossibility of knowing a quantum particle's position and momentum perfectly precisely—to prove the second law of thermodynamics[1] and resolve Maxwell's demon paradox.[2] Neither hope prevailed,[3] but both illustrated how thermodynamics sparks curiosity when superimposed on quantum theory.

In 1956, Harvard physicist Norman Ramsey showed that qubits could have temperatures below absolute zero.[4] We'll see in chapter 7 how such temperatures are possible. Not that Ramsey thought of qubits as qubits, since the quantum-information state of mind hadn't crystallized by his time. Rather, he described two rungs in an atom's energy ladder. A bunch of such atoms could serve as a quantum engine, Erich Schulz-DuBois and Henry Scovil realized three years later.[5] They detailed their vision with Joseph Geusic,[6] as we'll see in chapter 7. All three physicists worked at Bell Labs—as had Claude Shannon, the founder of information theory.

During the 1970s, quantum thermodynamics benefited from a crossbreed between mathematicians and physicists. Mathematical physicists developed equations that model how quantum systems settle down into equilibrium.[7–11] The 1980s advanced the study of quantum engines—theoretical tools for analyzing them and experimental tools for devising them. Ronnie Kosloff of the Hebrew University in Jerusalem and Robert Alicki of Gdańsk University in Poland adopted an abstract, mathematical approach.[12,13] For instance, Ronnie identified properties of a quantum system that enable it to serve as an engine, be the system an atom, a spin, or anything else.

Adopting a more concrete approach, Marlan Scully explored the thermodynamics of quantum systems such as lasers. I found Marlan's epithet, *the quantum cowboy*, apt when he visited the institute I belong to as a postdoc. Raised in Wyoming, Marlan is a professor and cattle rancher in Texas. He takes a tell-it-like-it-is approach to science and communication.*

Also during the 1980s, MIT theorists sought to construct a theory of quantum thermodynamics. The cohort included Elias Gyftopoulos and George Hatsopoulos, who recruited Gian Paolo Beretta. They reevaluated entropy, equilibrium, and tenets of thermodynamics, with small and out-of-equilibrium systems in mind. Some scientists dismissed the cohort's work: Thermodynamics had always focused on large systems. How could a thermodynamics of small systems not be an oxymoron? Four decades later, it's a subfield of science.

A few years later, MIT hired another cofounder of quantum thermodynamics: Seth Lloyd is the quantum-computing scientist

---

* Marlan and one of his sons have coauthored a quantum-thermodynamics book, *The Demon and the Quantum: From the Pythagorean Mystics to Maxwell's Demon and Quantum Mystery.*[14]

(chapter 3) who quips about the second law of thermodynamics (chapter 4). As I warned, if you dream up any idea about quantum computation, Seth likely wrote about it a few decades ago. The warning extends to quantum thermodynamics. Seth's 1988 PhD dissertation refers to Maxwell's demon in its title.[15] The thesis explores how one can and can't use information to lower thermodynamic entropy. The thesis also shows how, under special conditions, classical thermodynamics nearly predicts a quantum system's behavior—even though the system can be entangled.* Examples include black holes, the densest regions in the universe.[18]

Quantum physics, information theory, and thermodynamics combine in the study of black holes. Physicists including Jacob Bekenstein, Stephen Hawking, Bill Unruh, and Paul Davies explored this setting during the 1970s and 1980s. But the black-hole community has worked mostly separately from the quantum-thermodynamics community. Exceptions exist and are growing; examples include Seth Lloyd's thesis and a story we'll encounter in chapter 14.

We've seen how else the 1980s benefited quantum thermodynamics. First, Paul Benioff and others explored the limitations on how little heat a computer can dissipate. Their focus on fundamental limitations led to the development of quantum computing. Second, Charles Bennett resolved Maxwell's demon paradox in 1982 (chapter 5). What a decade.

The embers of quantum thermodynamics burned lower during the subsequent two decades. Among those blowing on the coals was Ilya Prigogine, the archaeology-studying, piano-playing Nobel laureate who explored nonequilibrium thermodynamics. Prigogine and collaborators reformulated quantum theory around the notions of irreversibility and time's arrow.[19]

Quantum thermodynamics—dare I say—picked up steam

* Related explanations emerged during the early 2000s.[16,17]

during the early 2010s. Quantum information theory had matured as a mathematical and conceptual toolkit, and it had begun shedding light on other fields. Scientists were understanding chemistry and materials anew, in terms of the information that quantum systems share and convey. Thermodynamics beckoned to quantum information theory, calling for reexamination. The combination roared into life over the past decade.

For much of the decade, quantum thermodynamics thrived mostly outside the United States. Europe, Canada, and Japan caught on early. A few hot spots burned beyond; for instance, Ronnie Kosloff had been pondering quantum engines in Israel for decades. Hot spots then emerged in Singapore, which had invested in quantum computing, and elsewhere. Only during the past few years has quantum thermodynamics gained traction in the United States. A handful of us, scattered across the country, waved the quantum-thermodynamics banner. The first quantum-thermodynamics conference to take place on American soil, to my knowledge, transpired in 2017. But quantum thermodynamics has been gaining popularity in the United States at last. Colleagues in other fields are borrowing our tools and inviting us to apply for funding together. Students and postdocs email me inquiries about opportunities to undertake research in quantum steampunk.

Why has the United States taken so long to catch on to quantum thermodynamics? The question confounds especially because roots of quantum thermodynamics grew in American soil during the 1980s. I haven't pursued the question with data and rigor, so I can't answer with authority. But a finding by a historian of science has given me pause. Much of quantum thermodynamics, especially early quantum thermodynamics, is foundational and theoretical. Quantum thermodynamicists have proved lemmata and theorems, scrutinized subtleties in probability theory, and reformulated the laws of thermodynamics. Even many proposers of quantum engines have had little intention of fabricating their ideas. The field's early foundational bent, the historian pointed out, resonates with the European philosophical tradition, which predates Socrates. In contrast, the United States has developed a tradition of innovation and practicality. Consequently, much of the country's science centers on experiments, technology, and applications. Little wonder that the United States began to welcome quantum thermodynamics only after the field grew beyond abstract theory.

I adore the abstract theory, to which I contribute. But I also work with experimentalists on ushering the theory into the real physical world. I also usher quantum thermodynamics outside its neighborhood into other disciplines, such as condensed matter, atomic and laser physics, chemistry, and black-hole physics. As quantum information theory began transforming other fields—including thermodynamics—during the early 2000s, so quantum thermodynamics is illuminating other fields anew. I feel fortunate to belong to a lively community of colleagues with whom I collaborate on all these aims.

Quantum Is Different

The twentieth-century physicist Philip W. Anderson coined the slogan "More is different."[20] He earned a Nobel Prize for elucidating statistical mechanics and condensed matter. His slogan encapsulates why we should bother studying statistical mechanics, the study of large, many-particle systems. Granted, every particle in a large system obeys Newton's laws or quantum theory. So, to describe large systems, we seem to need no theory beyond classical and quantum mechanics. But we'd apply Newton's laws to a cloud of steam by calculating every steam particle's trajectory. The steam cloud contains about 10^{24} particles; so, we'd spend oodles of time, energy, and computation. Worse, we'd gain little insight.

Many-particle systems exhibit collective behaviors off-limits to individual particles. If you only ever watch one raven, you can't fathom a flock—its undulations and roilings, its fabric-like qualities. More is different.

When it comes to heat and work, quantum is different. We characterized heat as the unharnessed energy of random motion. In contrast, work is organized energy available for performing tasks. We saw one example of quantum work in chapter 5, when quantum particles lifted a tiny teapot via Szilard's engine. But we might want general rules, beyond one example, for defining and measuring quantum work and heat. Furthermore, more subtleties than I let on obscure the work and heat exchanged by Szilard's engine.

To see why, imagine a variation on Szilard's engine, another quantum gas in a box. The gas exchanges heat with a heat bath through the box's walls. A piston compresses the gas, performing work on it. How can we define and measure the heat absorbed by the gas and the work performed on the gas? The gas is in some quantum state. In most quantum states, the gas lacks a well-defined energy, due to quantum uncertainty. If the gas lacks a well-defined energy now, and it lacks a well-defined energy after the piston and bath act, how do we put a number on the change in the gas's energy, let alone split that change into heat and work?

We could measure the gas's energy before and after the protocol. Each measurement would force the gas into a state with a well-defined energy. Viewed from another angle, however, this antidote serves as poison: Each measurement disturbs the gas's energy, changing the energy. Instead of having to tease apart only heat and work, we now have to tease apart heat, work, and the energy injected by the measurement. So how well can we port over, into quantum thermodynamics, the classical definitions of heat and work? About as well as we can port over, into statistical mechanics, the mindset behind Newton's laws.

What's a scientific community to do when it can't distinguish work from heat as it's used to? Everybody and their uncle proposes a definition of quantum work and heat. The definitions could fill a menagerie in the metropolis of ideas, as a testament to humans' imagination and argumentativeness. Many proposers believe that they've resolved the problem, so not everyone agrees on any one resolution.

I haven't proposed a definition. I believe that different definitions suit different contexts, so no one definition need rule them all. Physicists are famous for trying to unify theories—for instance, quantum theory, which governs small objects, with general relativity, which governs enormous objects. But the same approach—unification—doesn't necessarily suit every problem. Whenever I find a paper that introduces a new definition of quantum work and heat, I save it in a folder labeled "Menagerie—definitions of quantum heat and work." Images of Victorian menageries depict cages scarcely larger than their inhabitants. Lions and tigers lash their tails alongside monkeys and elephants; none occupies an environment that resembles its home. Therefore, I eschew the design of the Victorian menagerie. I imagine my specimens—the different definitions of quantum heat and work—as padding, climbing, and flitting around a Victorian botanical conservatory. Silver beams curve upward into elegant shapes through which you can glimpse the sky; the high, glass roof lends an airy atmosphere. Let's meet some of the denizens.

One definition of quantum work and heat, I envision as a Yorkicockasheepapoo—a mix of Yorkshire terrier, cocker spaniel, English sheepdog, and poodle. Nature wouldn't concoct such a combination; only a breeder would. Likewise, nature wouldn't suggest this definition of quantum work and heat; only a scientist—likely a theorist—would. We can use this definition if two conditions are met. First, only one thing ever happens at a time. Second, after anything happens, we measure our quantum system's energy.

For instance, consider the quantum gas in a box. The gas exchanges energy with a heat bath through the box's walls, and a piston compresses the gas. To define heat and work in Yorkicockasheepapoo fashion, we break the process into steps: we measure the gas's energy, then let the gas interact with the bath, then measure the gas's energy, then move the piston inward, and then repeat those steps. The energy reported by our detector after a heat-bath coupling, minus the energy reported before, we define as heat. The energy reported after a piston movement, minus the energy reported before, we define as work.

This definition of work and heat offers two advantages. First, we've defined the work and heat exchanged in every experimental trial—not only on average over trials, as in a later definition. Second, this definition is operational—expressed in terms of experimental steps. How to measure work and heat, using this definition, is relatively clear.

On the downside, each measurement disturbs the system's energy, as we've discussed. Also, the definition suffers from artificiality: Systems often exchange heat while absorbing work. Separating the steps is unnatural, as is measuring the energy frequently. So, this definition looks to me like a creature designed by breeders—a Yorkicockasheepapoo.*

Definition number two of quantum work and heat, I envision as an elephant—an animal that has a solid physical presence. This definition calls for a battery, an auxiliary system separate from the system of interest. We should be able to reliably deposit energy in, and extract energy from, the battery. For example, suppose that the system of interest is a tiny quantum spring owned by Audrey. The battery can be an atom supplied by Baxter (figure 6.1). The simplest quantum battery has only two energy levels. Baxter can keep his atom at a low energy, so that it occupies the lowest or second-lowest

* Disclaimer: I have used this definition, as a theorist.

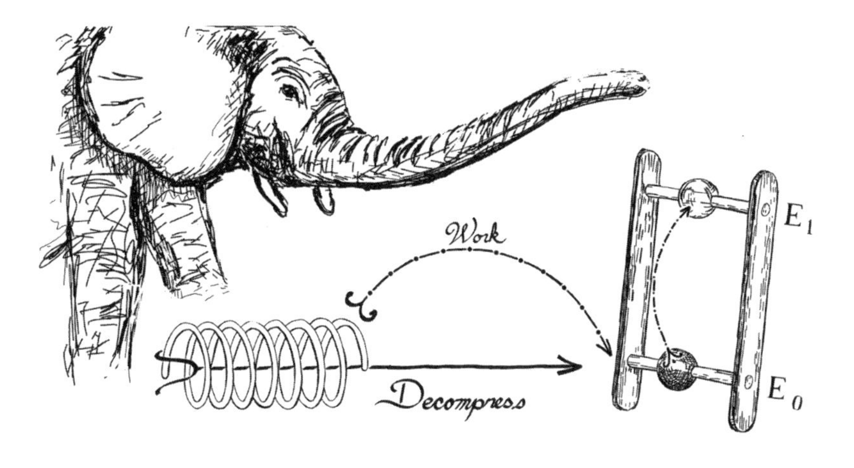

FIGURE 6.1

rung of its energy ladder. Let's assume, for now, that the battery always has a well-defined energy, E_0 or E_1.

Suppose that the siblings wish to extract work from a compressed spring. Baxter prepares his battery in its lowest level, with an amount E_0 of energy. The siblings couple the spring to the battery, enabling energy to slosh between the systems. The spring ends in a low-energy, relaxed state. The atom ends on its upper ladder rung, with an amount E_1 of energy. We define the work performed by the spring as the energy gained by the atom, $E_1 - E_0$.

Now, suppose that the siblings want to perform work on the system of interest—to compress the spring. Baxter prepares his battery on its upper ladder rung, with an amount E_1 of energy. The siblings use energy from the battery to exert a force on the spring. Providing the force depletes the battery's energy to E_0; we can say the spring has absorbed an amount $E_1 - E_0$ of work.

This definition offers three advantages. First, it prescribes a physical protocol that an experimentalist can perform to measure the work. The ability to measure a quantity, such as work, lends weight to its meaningfulness. Second, the protocol doesn't require us to measure the spring's energy directly. We therefore disturb the spring less than in the Yorkicockasheepapoo protocol. We do disturb the spring, however, by coupling it to the atom. But the Yorkicockasheepapoo protocol hits the spring's energy more like a lightning bolt.

Third, we can generalize the protocol: Suppose that you don't know precisely how much work you'll perform to compress the spring; you can only estimate. The work required might not equal the battery's energy gap, $E_1 - E_0$. If not, even if the work required is less than $E_1 - E_0$, you can't use the battery reliably. The atom typically can't pause between energy-ladder rungs, offering less work than you need. But we can devise a battery that has many closely spaced rungs. If you underestimate or overestimate the work required, no problem. The battery can end a little higher or lower than anticipated.

Two drawbacks mar the elephant definition. First, it prescribes a means of defining and measuring just work, not heat. Second, we can't always assume that the battery has a well-defined energy: The battery is a quantum system that can be in a superposition of energies.[21] If the battery is, how much work was performed remains unclear.

Definition three of quantum work and heat resembles a wildebeest that's neither fast nor slow, neither adventurous nor a straggler—the middle of the pack, or the average. Quantum physicists invoke averages all the time. Imagine running an experiment on some quantum particles in each of many trials. At the same time in each trial, we measure the particles' energy. Then, we average the outcomes. A theorist can predict this average, knowing the particles' quantum state and knowing properties of the particles' environment. As the state and the environment change across a trial, so does the particles' average energy. State-sourced changes to the average energy constitute heat, and environment-sourced changes to the average energy constitute work, according to the wildebeest definition.

Let's see why this definition makes sense, starting with the environment. By the particles' environment, I mean, are the particles in an electric field, in open space, or in a box? Is the box being pushed? Can the particles move only on a two-dimensional surface, like a tabletop? Does the tabletop slope like a hill? Experimentalists change the system's environment by turning knobs in the lab—by strengthening a magnetic field, turning on a laser, and so on. Experimentalists control these changes, just as a thermodynamic agent controls energy that's serving as work. So, we define as work any changes caused to the average energy by environmental changes.

The average energy depends, as well as on the environment, on the particles' quantum state. The state, we established, is the quantum analog of a probability distribution. Probabilities dictate how accurately we can predict how events will unfold. If the possible unfoldings have equal probabilities, the event is completely random, and we can't predict much about it. Heat, the energy of random motion, randomizes events. So, heat homogenizes probabilities and, likewise, homogenizes quantum states. We therefore define as heat any changes caused to the average energy by changes in the quantum state.

This wildebeest definition offers the advantage of according with intuition: work is controlled, and heat randomizes probabilities. But a quantum average involves many trials. The wildebeest definition doesn't define, or show how to measure, the work and heat exchanged in one trial.

Also, the wildebeest definition contradicts a definition favored by condensed-matter physicists. Condensed-matter physicists study matter. A chunk of matter has an energy ladder that contains many, many rungs. Imagine thwacking the chunk periodically for a long time: Wack! Wack! Wack! Wack! Condensed-matter physicists dignify this thwacking with the name *Floquet driving*, after the nineteenth-century French mathematician Gaston Floquet. Imagine measuring the matter's energy afterward. Our detector will report the number associated with some energy-ladder rung. The detector's probability of reporting *this* rung number equals the detector's probability of reporting *that* rung number, and so on. In other words, the matter's quantum state is spread evenly across all the energy rungs. The quantum state is spread the same way if the matter has equilibrated with an infinite-temperature bath.* So the thwacking—pardon me; the Floquet driving—heats the qubits, according to condensed-matter physicists. According to the wildebeest definition, the thwacking provides not heat but work. After all, the thwacking results from a change in the matter's environment, not from a heat bath. So condensed-matter physicists wrangle with quantum thermodynamicists about the wildebeest definition.

Let's train our binoculars on one more specimen in the menagerie of definitions for quantum heat and work. This definition reminds me of a hummingbird, which scarcely disturbs the twig it alights on. This definition stipulates that we measure the quantum system's energy weakly. To understand what a weak measurement is, we have to understand how quantum measurements operate.

Imagine the Stoqhardt siblings measuring a system of Audrey's, such as an atom. They couple the atom to some system of Baxter's; for instance, they lob a photon at the atom. The photon bounces off the atom, exchanging energy, momentum, and spin with the atom. The exchange correlates the photon's quantum state with the atom's quantum state.† Then, Baxter observes a property of his photon. For example, the photon hits a photodetector, a camera that collects light. The camera registers the photon's

* An infinite-temperature bath sounds extraordinarily hot. So, you might expect an infinite-temperature bath to boost the matter to its top energy level. The bath doesn't; although we'll see a bath that does in chapter 7.

† This use of the term *correlate* differs slightly from the use in chapter 3. According to chapter 3, if Audrey's particle entangles with Baxter's, measuring Audrey's particle can yield an outcome correlated with a measurement of Baxter's particle. Here, I'm saying that entanglement correlates Audrey's particle with Baxter's. Consider this second usage as shorthand for the first usage.

energy, disturbing the photon. The camera shifts a needle across a dial, till the needle points at a particular number. This process transduces information about the photon from the quantum scale to the human scale. Baxter reads the number off the dial, gaining information about the photon. From that information, and from the correlation between his and Audrey's systems, the siblings infer about Audrey's atom.

Physicists often assume that Baxter's system entangles maximally with Audrey's—that the systems correlate as strongly as possible. In this case, the siblings can infer about Audrey's atom as much information as any measurement can reveal about a quantum state. We call such a measurement of Audrey's atom *strong*. But the correlation can be weak: the photon can interact with the atom for a short time, without exchanging much energy, momentum, or spin. Baxter's photon will provide little information about Audrey's atom. Baxter will perform a *weak measurement*.

Why would the siblings forfeit information about the atom? To avoid disturbing the atom much. If the systems entangle maximally, the camera jolts the atom similarly to how it jolts the photon. If the systems entangle only a little, the camera disturbs the atom as a hummingbird disturbs you by hovering behind your left ear: a shiver might run down your spine, but you suffer no violence. Furthermore, the siblings can run many trials of their experiment. They'll accumulate many measurement outcomes. From those, the siblings can reconstruct information that they could have obtained from strong measurements, without disturbing the system's energy as much.

The hummingbird definition of quantum work and heat begins with the Yorkicockasheepapoo definition: measure the system's energy. But we replace the strong energy measurements with weak measurements. This definition offers the advantage of scarcely disturbing the system's energy. But drawing conclusions tends to require many trials, so this definition says little about the work and heat exchanged in one run.

Here ends our visit to the menagerie of definitions of quantum work and heat. We've met the Yorkicockasheepapoo (measure the energy frequently, and separate work exchanges from heat exchanges temporally), the elephant (define work in terms of a battery), the wildebeest (define work and heat in terms of an average—the middle of the pack), and the hummingbird (measure the energy weakly). Other species roam beyond the camellias and palms. But let us slip out, the glass door shutting behind us. Turning around for one more glimpse, we can judge the glinting edifice from outside. The menagerie signifies, to me, the trickiness of translating even basic thermodynamics into quantum theory. Quantum is different.

Now, let us confront the terrain in front of us. I envision quantum steampunk as a landscape. A map of it would consist of a parchment like the one examined by Audrey at the beginning of this chapter. Many city-states, principalities, and villages dot the map, as quantum steampunk encompasses many communities. Different communities address different corners of quantum thermodynamics and approach the subject from different angles. We'll meet the communities one by one, traversing the landscape. Audrey, Baxter, and Caspian will traverse their map at the same time. We won't visit every grotto and ruined castle, and the ones we visit will be ones that your humble tour guide happens to know. But we'll develop a sense of the scenery through stories. So, pack your trunk, fetch your binoculars, and secret a penknife in an inner pocket. As Caspian said, there be dragons all over the map.

CHAPTER 7

PEDAL TO THE METAL

QUANTUM THERMAL MACHINES

Put-put-put . . . put PONK. The train juddered several times, and then slowed to a stop. Baxter took advantage of the confusion to swap one of his cards for a fresh one from the stack. Audrey slapped his hand, as a young conductor stumbled down the aisle, his navy-and-gold cap askew.

"Excuse me, sir," she called out. "What has happened?"

The conductor grabbed the brass rail across from her seat and steadied himself on it while straightening his cap.

"Sounds like one of the engines 'as cut out, miss," said the conductor, wiping his glistening forehead.

Audrey exchanged a glance with Caspian, who raised an eyebrow from above his newspaper. Baxter swapped another card for the top of the stack.

"Not to worry, miss," the conductor added, wiping off another bead of sweat. "We've reached the outskirts of Manchester, which 'as more quantum engineers than you could spit at—beggin' your pardon, miss. We'll be back on our way by evening, mark my words.

THERMAL MACHINES are devices that use, produce, or store heat or work. Examples include heat engines, refrigerators, heat pumps, ratchets, batteries, and clocks. The thermal machines available for purchase today obey classical physics. But what if they contained quantum components?

Physicists Erich Schulz-DuBois and Henry Scovil proposed the first quantum engine, as mentioned in chapter 6. They extended their 1959 proposal[1] with physicist Joseph Geusic in 1967.[2] We'll begin with their vision, as the field did; although understanding their engine doesn't require information theory. Then, we'll listen to the purrs of quantum engines designed after theirs. We'll also contemplate the smallest possible refrigerator and how entanglement can benefit batteries. The field of quantum thermal machines has helped us understand how quantum is different, but you won't find quantum engines on the road anytime soon.

{ VROOOOOM! }

The quantum engine to be imagined consists of a variation on a laser. Perhaps you've used a laser pointer during a PowerPoint presentation; or admonished someone to "be *careful* with that; you almost shone it in my eyes!"; or teased your cat by bouncing a laser light around so that it resembled a small, catchable creature. We can form a laser by putting a bunch of atoms in—you guessed it—a box. Each atom has the same energy ladder as every other. *Lasing*—a laser's action—involves three rungs, whose energies we'll call E_0, E_1, and E_2. The atoms climb and descend the rungs while absorbing and emitting photons. The emitted photons form the light that drives cats mad.

A maser resembles a laser but emits microwaves, rather than visible light. The maser predated the laser, but no one needs microwaves during a PowerPoint presentation (except any audience members wanting popcorn during the show). Guess where the maser and laser were developed? Bell Labs—home of Claude Shannon, Schulz-DuBois, Scovil, and Geusic—as well as at the Lebedev Physical Institute in Moscow. The first quantum heat engine we'll see consists of atoms that can form a maser, although the atoms will behave a little differently than when acting as a maser.

The engine interacts with three heat baths, all at different temperatures. Two baths exchange photons with the engine. One bath

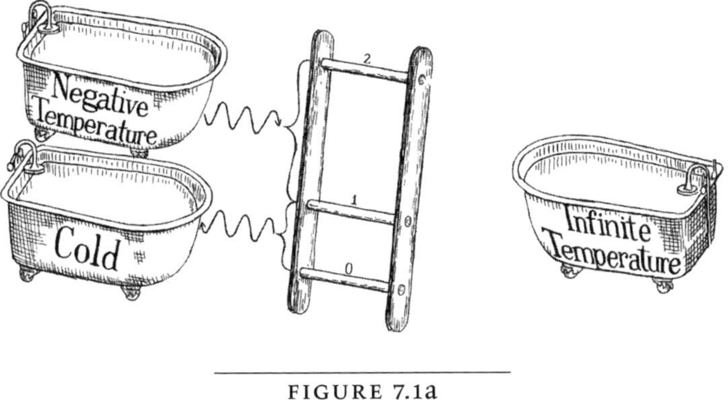

FIGURE 7.1a

FIGURE 7.1b

can't necessarily exchange photons, for technical reasons. But we can still envision all the baths as exchanging with the engine packets of fixed amounts of energy.

One bath emits photons of energy $E_1 - E_0$, which can boost the engine from rung 0 to rung 1 (figure 7.1a). Alternatively, the engine can fall from rung 1 to rung 0 while emitting photons of energy $E_1 - E_0$ into that bath. The second bath exchanges packets of energy $E_2 - E_1$ with the engine, and the final bath exchanges photons of energy $E_2 - E_0$ (figure 7.1b). So, each bath facilitates atomic transitions between two energy rungs. The atom "looks" to each bath as though it has only two rungs. Those two rungs serve similarly to the spin-up and spin-down states of chapter 3: each pair of energy rungs forms a qubit.

What happens when a qubit, formed from two energy rungs, equilibrates with a bath? The bath decoheres the qubit, robbing it of quantum properties other than energy quantization. For instance, the qubit can no longer be in a superposition, so the qubit has some well-defined amount of energy. But we don't know how much energy if we only know that the qubit is at thermal equilibrium with the bath. The qubit has some probability $1 - p$ of occupying its lower energy rung and a probability p of occupying its upper rung. The probability p depends on the bath's temperature: the hotter the bath, the greater the qubit's probability of occupying the upper rung.

How does the probability p depend on the bath's temperature? We'll find out by considering a low temperature, checking how the probability behaves, imagining raising the temperature, seeing how the probability changes, and repeating. Let's start with the lowest possible temperature. According to the third law of thermodynamics, no system's temperature can reach absolute zero. But a temperature can almost reach absolute zero; so, let's pretend that the bath's temperature does reach absolute zero. The qubit will have no energy; the upper energy rung will lie out of reach. Therefore, the qubit's probability p of occupying that rung will be zero.

Imagine raising the temperature slightly above zero. The qubit acquires a tiny probability of occupying its upper rung. As the temperature grows, the probability grows, but much more slowly than the temperature. I think of the temperature as an extrovert and the probability as an introvert: A speech by the extrovert coaxes a sentence from the introvert; laughter by the extrovert secures a smile from the introvert. Likewise, a leap in the temperature accompanies a baby step by the upper rung's probability.

This behavior climaxes when the temperature reaches infinity: the upper rung's probability reaches one-half. Not even the dizzying heights of infinite temperature can guarantee that the qubit occupies its upper energy rung. This result may surprise us, but it's how the mathematics works out.

Yet temperatures of another type can push a qubit onto its top energy rung: temperatures below absolute zero. Before we address the question, "What in tarnation are temperatures below absolute zero?" allow me to conclude my conceit. I am the qubit's probability of occupying its upper level; I am an introvert. I delight in quiet contemplation, in the company of a few, cherished friends. Extroverts try to draw me out, but their efforts can counteract their aims: The more they talk, the less silence I have in which to reflect, form opinions, and speak. The higher they climb the staircase of gregariousness, the more they leave me behind; I remain at one-half while they reach infinity. To those kind extroverts seeking to engage us introverts, I suggest a strategy gleaned from thermodynamics: Descend from the heights. Return to the finite realm, and replace your effusiveness with its negative—listening. It will boost us introverts to a probability p of one.

Now, what in tarnation are temperatures below absolute zero? I'm measuring temperature in units of Kelvin, introduced in chapter 4, rather than in degrees Fahrenheit or degrees Celsius. Zero Kelvin, or absolute zero, is the lowest temperature toward which one can hope to cool any system. So, temperatures below zero—negative temperatures—seem impossible. But a negative-temperature qubit isn't colder than a zero-temperature qubit; it's hotter than an infinite-temperature qubit. At infinite temperature, we said, a qubit has a 50% chance of occupying its upper energy rung. Imagine setting the temperature to infinity, then pumping more energy into the qubit. We can pump energy into an atom by shining a laser at it. The qubit will acquire a greater-than-50% probability of occupying its upper rung. The qubit's temperature can't lie between zero and infinity, we've seen. The temperature turns out to be negative, according to the mathematics.

So a negative-temperature qubit has more energy—is hotter—than an infinite-temperature qubit. That's why temperature can dip below zero, the temperature of the coldest system conceivable. Negative temperatures make systems much hotter than zero temperature.

You and I can't have negative temperatures. Neither can the air outside; nor your oven; nor the water from that tap, in the first-floor bathroom in your office building, that always comes out scalding. Each of these systems has an energy ladder with infinitely many rungs. A qubit has an energy ladder with only two rungs. The finite number of rungs allows for negative temperatures. The reason why lies beyond the scope of this book. Suffice it to say that quantization allows quantum systems to have negative temperatures, that negative-temperature systems are hotter than infinite-temperature systems, and that technologies used today—including lasers—leverage negative temperatures.

One of the heat baths interacting with our engine has a negative temperature. We might as well think of this bath as consisting of negative-temperature qubits. It gains or loses energy as the atom transitions between the top two energy rungs, 1 and 2 (figure 7.1a). The atom more likely resides on the top rung, due to the negative temperature. The second bath—the *cold bath*—facilitates transitions between the bottom two rungs, 0 and 1. This bath has a temperature slightly above zero. So, the bath gives the atom a decent probability of occupying its lowest rung.

The third heat bath helps the atom transition between its top and bottom rungs, 2 and 0. This bath has an infinite temperature, and it exchanges *work* with the engine. How can it exchange work if heat baths exchange *heat* and I fussed about distinguishing heat from work in chapter 6? You can say about some people that "their word is as good as gold." You can say about infinite-temperature heat baths that "heat exchanged with them is as good as work." The reason is, exchanging heat doesn't raise such a bath's entropy. Imagine funneling heat from an infinite-temperature bath to a Carnot engine. The engine will transform all the heat into work, without dissipating any. We might as well regard any heat exchanged with the bath as work.

Figure 7.1 depicts how the quantum engine functions. Each energy rung, in the figure, carries a little ball. The ball represents the atom's probability of occupying that rung in a given trial, the rung's *probability weight*. The greater the probability weight, the larger the ball.

The cold bath drops some of the probability weight from rung 1 to rung 0. Meanwhile, the negative-temperature bath boosts some of the probability weight from rung 1 to rung 2. The negative-temperature bath is so hot that it boosts more probability weight than the cold bath drops. Therefore, much probability weight occupies the ladder's top rung, some occupies the bottom rung, and hardly any occupies the middle rung.

Now, the atom equilibrates with the infinite-temperature bath (figure 7.1b). So, the infinite-temperature bath coaxes the atom into having half its probability weight on rung 2 and half its probability weight on rung 0. So, some probability weight drops from the top rung to the bottom. In other words, the atom drops from rung 2 to rung 0 in some trials. During the drop, the atom emits a packet of energy into the infinite-temperature bath. So, the quantum engine performs work.

Geusic, Schulz-DuBois, and Scovil calculated the engine's efficiency—the bang that the engine gets for a buck, or the work performed per packet of heat absorbed from the hot bath. The negative-temperature bath gives the atom an amount $E_2 - E_1$ of heat. Afterward, the engine gives the infinite-temperature bath an amount $E_2 - E_0$ of work. The work exceeds the heat, as figure 7.1 illustrates. The quantum engine operates at an efficiency greater than one; the atom gets, per buck, more than a buck's worth of bang. The engine might as well have visited the store with enough money for a can of oil (not that engines buy their own oil) and left with two cans, due to a buy-one-get-one-free deal. The hot bath's negative temperature, with the cold bath's low temperature, subsidizes the deal.

How quantum is this engine? The atom has quantized energies, so it's quantum literally. But Geusic and colleagues never leverage entanglement, nor even any superpositions. Classical systems can approximate the behaviors reported by Geusic and colleagues. But the atom, endowed by quantum physics with few energy levels, forms a natural platform for such an engine. Furthermore, Geusic and colleagues didn't explore all the behaviors that the engine can possibly exhibit. Some, such as the engine's reliability, rely on wavelike properties of quantum systems.[3]

Ladies and Gentlemen, Start Your Quantum Engines

Geusic, Schulz-DuBois, and Scovil did to the field of quantum engines what a green flag does to a NASCAR race. The field of quantum engines didn't rev up as quickly as a race car, but results have piled up by now. Many results center on the Carnot efficiency, the greatest efficiency of any engine that contacts only two heat baths. An engine can achieve the Carnot efficiency only by operating infinitely slowly, for a reason that brings to my mind the 1954 film *On the Waterfront*. In the movie, a former boxer laments his lost shot at greatness with the famous line, "I coulda been a contender." Similarly, at the end of a quick engine cycle, part of the heat dissipated can lament, "I coulda been work"—albeit without actor Marlon Brando's sulk or Brooklyn accent. To avoid dissipating energy, you have to reduce the engine's power to zero. The *power* is the rate at which the engine performs work. Power trades off with efficiency in classical thermodynamics.

A quantum engine may be able to have its power and eat its efficiency, too. Italian physicists Michele Campisi and Rosario Fazio envisioned an engine formed from particles that can undergo a special phase transition.[4] You've witnessed a phase transition if you've boiled water for tea, transforming water into steam. Quantum systems can undergo phase transitions more exotic than the ones in our everyday lives. If you drive a certain quantum system across a certain phase transition, interactions between the particles may boost the work performed by the engine. The power will grow without denting the efficiency.

One might hope that a many-particle quantum engine would resemble a car engine, miniaturized. Tiny pistons would pump, turning tiny gears. Alas, I can offer no such romantic vision. The engine might consist of a glob of particles, a miniature cloud hovering above a tabletop as though threatening snow. Few systems are known to undergo the necessary phase transition, which is exotic even to quantum physicists. Campisi and Fazio found evidence that at least one known system can undergo the transition, however. I hope that their proposal spurs experimentalists to identify more such systems—that a quantum-thermodynamic technology motivates fundamental discoveries.

Audrey alluded to another quantum engine in this book's prologue: "'Baxter is developing a superior [spy fly], which can extract energy from a certain type of light quite efficiently. Not the type here,' she added, waving at the fireplace, 'but a type that Ewart would have in his lab.'" Audrey is referring to *squeezed* light, which we can understand through the uncertainty principle. The uncertainty principle limits the extent to which a quantum system can have a well-defined position *and* a well-defined momentum. Light has two properties, analogous to position and momentum, that also obey the uncertainty principle. You can squeeze most of the uncertainty out of one property, leaving the other property monumentally uncertain.* Squeezed states facilitate applications of quantum information science, including metrology and cryptography.

We've seen engine cycles that involve a cold bath and a hot bath. Imagine replacing the hot bath with squeezed light. The light isn't at equilibrium, but we can attribute a temperature to it anyway.[5] Imagine running a quantum engine between the baths at maximum power: the engine performs the greatest possible amount of work per cycle. Running at maximum power erodes an engine's efficiency, we've seen. We've established also that, if an engine contacts just two thermal baths, the efficiency can't exceed the Carnot efficiency. But the squeezed-light engine runs at a higher efficiency than Carnot's, according to a 2014 analysis.[6]

Such a snub of Carnot shouldn't discomfit us: The squeezed bath isn't at equilibrium. It violates an assumption behind Carnot's bound on efficiency, so the engine has no obligation to obey the bound. But, before encountering the engine in the scientific literature, I wouldn't necessarily have predicted that squeezed light would break Carnot's bound. Edison might as well have predicted, upon beginning his experiments, which materials would support practical light bulbs.

Not everyone agrees that the squeezed-bath engine breaks Carnot's bound.[7,8] Theorists have argued that the squeezed bath transfers not only heat, but also work to the engine. We should deduct this work from the work performed by the engine, the theorists say. Quantum work and heat stir up controversy again.

Regardless of the debate, Zürich experimentalists fabricated

* Did we discuss squeezing in chapter 2 under a different name? According to chapter 2, measuring one property eliminates its uncertainty while boosting the other property's uncertainty. This measurement differs from squeezing: The measurement changes a quantum state abruptly and violently, as a crash reduces a moving car's speed. Squeezing changes the state as the acceleration pedal changes a car's speed: more gently, over an extended time. You can accelerate a car, using a pedal, to 30 miles per hour, or 45, or 50. Similarly, squeezing lets you choose how much uncertainty to cram into one property of light. Finally, measuring a particle can give its position a well-defined value for only a short time. Light stays squeezed for longer.

an engine that exchanges energy with a squeezed-light bath.[9] Their engine consists of a tiny metal bridge that connects two vertical posts. The bridge has the thickness of about a hundred DNA strands. Like a plucked violin string, the bridge can vibrate up and down. The greater the vibration frequency, the more energy the bridge has. The energy is quantized (one of just a few numbers, with gaps between them), so the frequency is quantized. I wonder how Brahms's violin concerto would sound if Stradivarius violins had similar gaps in their frequency ranges.

The bridge engine performs work—it outputs useful energy—when dropping from a high energy-ladder rung to a lower one. The work is performed on the posts that support the bridge. The experimentalists didn't bother channeling the work to perform any useful task, like raising a tiny teapot. But they could have, by clamping a to-be-worked-on device to a post.

How much work does the bridge perform per cycle, on average over experimental trials? The amount of energy carried by an infrared photon. An incandescent light bulb radiates about 10^{20} photons per second—and many of those carry more energy, apiece, than an infrared photon. Don't expect the squeezed-bath engine to revolutionize the energy industry tomorrow. Baxter must have advanced beyond today's experimentalists, to power a spy fly with a squeezed-bath engine. Still, I couldn't cool a nanobridge till it exhibited quantum behaviors, control the bridge, and measure the work it performed. So, today's experimentalists deserve applause, no less than a violin concerto.

Obsession with the Second Law

We've now seen three quantum responses to Carnot's efficiency and Carnot's engine cycle: the maser engine, the phase-transition engine, and the squeezed-light engine. Thinkers other than Carnot devised other engine cycles—other sets of steps that an engine can undergo to output work, then return to its initial conditions.

For instance, car engines undergo Otto cycles, named for the nineteenth-century German engineer Nicolaus Otto. My first foray into quantum Otto engines began over coffee while I was pursuing my PhD.

"Hey, you're interested in breaking the second law of thermodynamics, right?"

I looked up from pouring milk into my mug. Gil Refael stood in front of me, holding a latte. Gil is a physics professor and a condensed-matter theorist at Caltech. Like the other condensed-matter theorists, he has an office in a low, beige, Spanish-style building called Bridge. I used to see Gil in Bridge occasionally, but more often I saw him at Caltech's Red Door Café.

Gil's question piqued my curiosity. I don't expect that physical systems can break the second law of thermodynamics, although I expect that they can bend around it. And I have an interest in the second law, the way Beethoven had a soft spot for the piano.

"I'm obsessed with the second law," I said, putting down the milk jug.

"Could you break the second law," Gil asked hopefully, "with many-body localization?"

Many-body localization is a behavior exhibited by certain types of quantum matter, or large collections of quantum particles. The matter could consist, for instance, of a thousand atoms that repel each other. Such atoms occupy some landscape, which experimentalists can sculpt using lasers. Picture a rugged, random landscape, full of mountains that peak high up and valleys that dip low down. Suppose that an experimentalist measures the atoms' positions. The detector might suggest that one atom sits halfway up a hill; one, on a plateau; another, near a mountaintop; and so on. Loosely speaking, each atom is associated with a wave that peaks sharply at one spot.

Imagine waiting a little while after the measurement. If the atoms were typical quantum particles, the wave peaks would soften, and the waves would spread out across space. We would no longer be able to associate any atom with any one spot. But many-body-localized atoms behave differently: their waves remain tightly peaked for a long time after the measurement. So, we can think of the atoms as mostly sitting in single spots—as *localized*. Since the system contains many atoms, which interact with each other, we call the atoms a *many-body* system. Hence the name *many-body localization*, which characterizes a phase of quantum matter.

Many-body localization contrasts with two behaviors familiar to us. First, imagine that those atoms were tennis balls in Griffith Park, a hiking spot near Caltech. Any ball not on a plateau or atop a hill would roll downward, into a valley. Classical particles wouldn't remain where the quantum particles can.

Second, picture our favorite example of an equilibrating system, a classical gas in a box. Imagine measuring the particles' positions and finding the particles clumped together in one corner. Shortly thereafter, the particles will spread across the box. The second law of thermodynamics ordains this spread. As the particles disperse, the number of positions accessible to them grows, so the number of microstates accessible grows, so the gas's entropy grows. Resisting the urge to disperse, many-body-localized systems resist the second law. They don't break it: if you wait long enough, the atoms *will* spread across the system. But they're like a dachshund who refuses to leave a bush before sniffing every twig and digging at every root—in no hurry.

Gil and colleagues had studied many-body localization for years. They'd detailed properties of it, simulated it on computers, constructed protocols for realizing it experimentally, and initiated experiments. Which merits a pat on the back and a gold star, as physicists aim to uncover and understand properties of our universe. But Gil remained unsatisfied.

"What's many-body localization good for?" he asked me.

Physicists had proposed one application of many-body localization to technology: quantum memories. Imagine encoding information in one of the thousand atoms. Without localization, the atom would hop around the landscape and interact with its neighbors. Those would interact with *their* neighbors, which would interact with *their* neighbors. The information would spread across the many-atom system via entanglement. We've said of entanglement that the whole is greater than the sum of its parts. So, you couldn't retrieve your information by measuring the first atom, or even by measuring clumps of atoms and combining your measurements' outputs. You'd have to measure all the atoms together in a special way. Good luck; you might as well measure all the bricks in London simultaneously. But, under many-body localization, atoms travel slowly, and entanglement spreads slowly. You can recover information stored in such a system. Many-body-localized systems might serve, therefore, as quantum memories for quantum computers.

"Fine," said Gil. We knew of one application for many-body localization; but surely, we could invent more. Surely, we could invent thermodynamic applications, by leveraging the phase's resistance to the second law.

A few weeks, meetings, and cups of coffee later, I proposed a design for an engine that can occupy a many-body-localized phase. We improved and analyzed the design with two collaborators: Christopher D. White, who was pursuing a PhD alongside me, and Sarang Gopalakrishnan, then a postdoctoral research fellow.[10] Christopher specializes in simulating quantum systems on classical computers. Sarang specializes in everything that involves many quantum particles, as well as in a blend of snideness, dark humor, and resignation that's oddly endearing. Since physicists abbreviate *many-body-localized* as *MBL*, Gil dubbed our engine the *MBL-mobile.*

To explain how the MBL-mobile works, I should explain the opposite of many-body localization. Particles localize in the jagged landscape depicted at the bottom of figure 7.2. Let's progress clockwise around the figure. On its left-hand side, the lasers are adjusted so that the peaks descend, becoming small bumps, and valleys

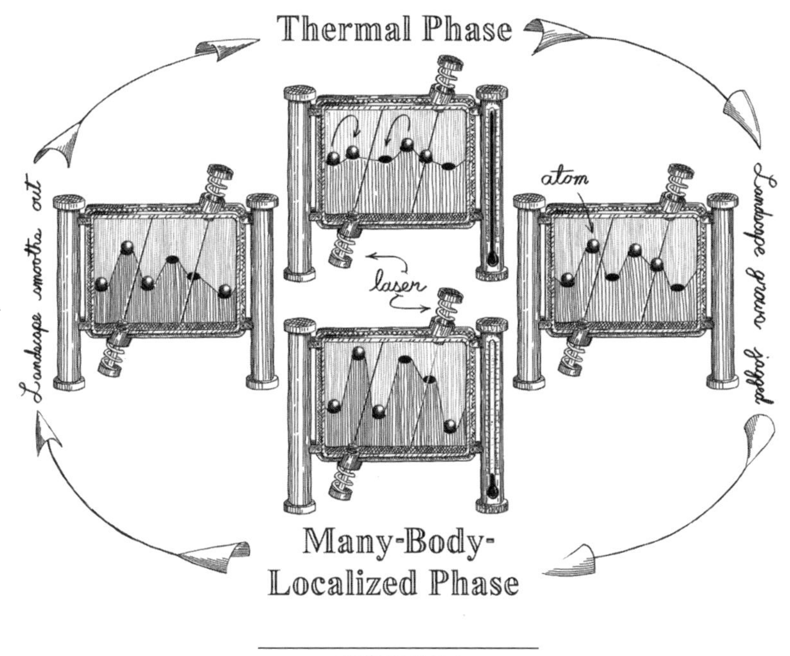

FIGURE 7.2

rise, becoming shallow dips. The atoms can hop around and interact easily, as in the top part of the figure. Quantum information spreads quickly; the system behaves similarly to the gas equilibrating in its box. The atoms occupy a phase of matter that we'll call the *thermal phase*. We can transition the atoms between their thermal and many-body-localized phases by adjusting the lasers. This transition resembles the transition of water into ice via a freezer.

One version of the MBL-mobile consists of about 10 atoms spread across a laser-induced landscape. The atoms undergo a quantum version of the Otto cycle, the engine cycle that powers cars. The engine begins the cycle in its thermal phase, in a mild landscape (top). The engine has a high temperature, being connected to a hot bath, which could consist of photons. We disconnect the engine from the bath and initiate the cycle's four steps.

First, we adjust the lasers to change the landscape from gentle to jagged (right-hand side). The engine transitions from its thermal phase to many-body-localized. Second, we connect the engine to a cold heat bath (bottom). The engine spews heat into the bath, dropping to a lower rung on its energy ladder. Third, we disconnect the engine from the bath. We then reverse our change in the lasers, filing down the landscape's jags (left-hand side). The engine returns from localized to its thermal phase. Fourth, we reconnect the engine to the hot bath, which replenishes the atoms with heat (top).

The MBL-mobile performs work, during the average trial, whenever transitioning between phases. The reason stems from a contrast between the many-body-localized and thermal phases: The set of atoms has an energy ladder whose rungs shift as the atoms transition between phases. This shift resembles the shift in the piston against which a classical gas expands in the classical Carnot cycle.

My collaborators and I didn't design a battery that would store the work performed by the engine. But you can stick the atoms in (of course) a box, between whose walls photons bounce back and forth. The engine could perform work on the photons.

So the contrast between many-body localization and the thermal phase enables our engine to perform work. The contrast also offers two more benefits. If you run an engine in many trials, you may encounter two dilemmas. Both come from engines' reliance on heat, energy whose randomness makes the engine a tad unpredictable. First, the amount of work performed may vary from trial to trial; the engine may be unreliable. Second, the engine may consume work, rather than perform work, in the occasional trial. Such work-consuming trials are undesirable, like putting laundry in a washing machine and pulling your clothes out dirtier than they went in. A many-particle quantum engine can suffer both shortcomings if cycled between two equally rugged landscapes. Cycling the engine between a rugged landscape and a milder landscape mitigates both problems: the engine grows more reliable and is less likely to consume work. The MBL-mobile benefits threefold from localization's distinction from equilibrium.

Sarang devised another way for many-body localization to benefit our engine. Ten atoms can't perform much work. What if we want to run a larger engine? We should manufacture many copies of the 10-atom mini-engine and run them in parallel. Transition the atoms not between phases, but between the localized phase's midlands and border: Start the engine not in its thermal phase, but shallowly localized; the atoms should move a little, but not too quickly. The landscape will begin the cycle with medium-sized jags, rather than being smooth. When we change the jags via the lasers, the atoms' energy-ladder rungs won't shift much; but they'll shift enough.

The atoms stay somewhat localized throughout the engine cycle.* Consequently, no particles from any mini-engine will stray into its neighbors. We can therefore pack the mini-engines close together, without worrying that one will interfere much with the others. The close packing makes this composite engine output much work per unit volume; localization improves the composite engine's power density. So, the composite engine can assume any size we please, from nanoscale to macroscopic.

How much of a punch could an MBL-mobile pack? Suppose that the hopping particles are electrons on a silicon surface studded with phosphorus atoms.† Our engine outputs less power than a car engine, so don't expect MBL-mobiles to infiltrate your garage next year. A quantum engine powers the train ridden by Audrey at the beginning of this chapter, but she lives in a steampunk novel.

---

* The atoms' staying put enables many-body-localized systems to store information as quantum memories, loosely speaking. So, the property of many-body localization that facilitates information storage enhances the engine, although the engine doesn't rely directly on information processing.

---

† Experimentalists had coaxed such a system not into many-body localization, by the time we wrote our paper, but into a similar phase.

The MBL-mobile holds its own, though: a car engine's power density exceeds our engine's by only a factor of ten or so.

The MBL-mobile merits comparison not only to car engines but also to another small engine: Some bacteria move by propelling themselves with flagella. A flagellum rotates due to a tiny motor in the bacterium. The MBL-mobile outputs more power than a flagellar motor by a factor of ten or so, according to our rough estimates.

Does the MBL-mobile break the second law of thermodynamics, as Gil hoped it might? Alas not; I know of nothing that breaks the second law. But the MBL-mobile leverages a quantum phenomenon—many-body localization—that resists thermalization for long times, caving to the second law only slowly.

As car models face crash tests, the MBL-mobile is facing an experimental test. Collaborators and I are working to manufacture an MBL-mobile from superconducting qubits. Fingers crossed that, in a later edition of this book, I'll report that the theory passed muster.

The MBL-mobile tickles my fancy not only because it works and, compared to certain competitors, works decently. The engine also fuses quantum thermodynamics with condensed matter, enabling each field to shed new light on the other. I enjoyed moonlighting in Bridge during the project. I retained an office with the quantum-computing group, to which I belonged most. But I felt a spark of pride when Gil introduced me to a visitor as a condensed-matter theorist. How apt is the name Bridge for the site of a bridge between disciplines.

{ The Energizer Bunny Meets Quantum Physics }

Quantum thermal machines extend beyond heat engines, which transform heat into work. Reversing a heat engine produces a refrigerator, for instance. My refrigerator draws power from a wall socket to cool the ratatouille left over from dinner. More generally, a refrigerator consumes work to transfer heat from a colder system to a hotter system. Collaborators and I imagined reversing the MBL-mobile's cycle, to pump heat from the cold bath to the hot bath. The baths—consisting of, for example, photons—are quantum. Systems tend to need to have low temperatures to exhibit quantum properties. So, the MBL-mobile could serve as a quantum refrigerator for the quantum equivalent of ratatouille.

The MBL-mobile contains many particles. How small can we shrink a refrigerator? Imagine a refrigerator intended to cool a particle of Audrey's. The refrigerator could run on work supplied to it from, for example, a wall socket. Alternatively, we could give the refrigerator a hot bath and a cold bath. The refrigerator could extract its own work from the two baths, then use that work to cool Audrey's particle. Scientists identified the smallest such refrigerator: a three-level system, such as the bottom three rungs of an atom's energy ladder.[11,12]

As refrigerators consume work and engines produce work, batteries store energy usable to perform work. We encountered a qubit battery in chapter 6; it contributed to the elephant in the menagerie of definitions of quantum work and heat. A qubit battery has two energy-ladder rungs analogous to an electron's spin-up and spin-down states. The battery is empty when occupying its lower energy rung. When occupying its upper level, the battery is full.

Imagine Audrey handing Baxter a qubit battery. It shares no entanglement with anything else but is in an otherwise arbitrary quantum state. Audrey directs Baxter to charge the battery. How well he obeys depends on two factors: how much energy he injects into the battery and how much time the charging takes. The more energy, and the less time, the better. Baxter aims to maximize the energy divided by the time—the power.

Since the battery is a qubit, we can represent its initial state with an arrow that points in some direction. Baxter chooses the battery's final state—represented by another direction—and the path from the initial state to the final. He can steer his arrow along the prime meridian, the equator, a spiral, a zigzag, or any other path. One path produces the greatest power.

Now, suppose that Audrey hands Baxter a set of qubit batteries. No battery is entangled with anything else, and Baxter must return the batteries unentangled. He must charge the batteries, maximizing the power received by the average qubit. Baxter could charge the batteries in parallel, applying his one-qubit strategy to each battery. But this approach wouldn't maximize the power per battery.

Baxter should entangle the batteries, and then later unentangle them.[13] This strategy boosts the power per battery by lowering the charging time, as follows. The set of batteries is in some quantum state. We can't represent that state by an arrow pointing in some direction in a three-dimensional space because the set of batteries isn't a qubit. But we can still imagine Baxter steering the batteries from one state to another, to another. If he couldn't entangle the states, he'd operate like a child who treads only on dry patches of sidewalk after a rainstorm: Baxter could follow one of just a few possible paths, which would zigzag and take him a long time. A child who doesn't mind wetting their boots can speed straight down the sidewalk. Similarly, if Baxter entangles the batteries, he can reach the final quantum state more quickly.

So, entanglement lets us charge quantum batteries quickly—at a high power. If only those recharging stations at airports could take advantage of entanglement! One traveler could recharge their phone, laptop, and tablet before their plane boarded.

But everyday batteries don't entangle. Nor will quantum thermal machines penetrate our daily lives in the foreseeable future. Experimentalists are building quantum thermal machines, but in proof-of-principle tests. Furthermore, whether the principle is worth proving isn't always clear. For instance, I failed in my first attempt at talking an experimentalist into performing an experiment. He operated an atomic lab, and I tried to convince him to measure the work performed by a quantum system.

"Why bother?" he asked.

A quantum system will provide hardly any energy. He could extract all the energy he liked from an outlet in a wall.

Fair point. Worse, the experimentalist would have drawn more energy, to cool his atoms and run the engine, than he'd have received from the engine. Quantum thermal machines have yet to benefit technology as the steam engine did.

Autonomous quantum thermal machines have been proposed to help resolve this challenge. Experimentalists invest work to run an engine when, for example, adjusting the MBL-mobile's landscape or strengthening a magnetic field. Autonomous machines run by themselves and so require no such work investment. Designs, abstract and concrete, exist; and one autonomous machine has been realized experimentally.[14] But autonomous quantum thermal machines have yet to show potential for transforming the energy market. Furthermore, although we needn't invest work to run autonomous machines, we invest work to cool them to near absolute-zero temperature.

Alternatively, molecular motors and nanorobots do operate in experiments and in organisms. For instance, proteins transport molecules along nanohighways in your cells. Three chemists won a Nobel Prize in 2016 for designing and crafting molecular motors. But such motors tend to follow the laws of classical statistical mechanics; quantum thermodynamics tends to be unnecessary.

Quantum thermal machines haven't sparked a Quantum Industrial Revolution. Nor have we broken the second law of thermodynamics, as some have hoped that quantum machines would. Still, we can bend around the second law, as a dancer bends beneath a limbo stick. And we've been pinpointing which achievements quantum machines can reach and classical machines can't, and distinguishing what's possible from what's impossible. So, quantum thermal machines have contributed to fundamental physics. What they contribute to engineering, we have yet to establish.

CHAPTER 8

Tick Tock

QUANTUM CLOCKS

"We shall be late," Audrey said. She was pacing up and down the corner of Market Square, not registering the cobblestones, the barred doors of the shops that lined the square, or the predawn chill that suffused the air. Her attention remained fixed on the silver pocket watch cupped in her hands.

"I told Captain Okoli that we would meet him at the docks at five ten," she continued.

"Audrey," said Baxter.

"The hour is already eighteen till five, and we still have not the faintest idea in which direction the docks lie," she said, as a ragged-looking mutt sporting a torn ear padded up to the far end of the path she was pacing. The dog sat down on the cobblestones and whined, but her pacing ended half a yard in front of him, and she turned on her heel without noticing him.

"Audrey."

"If we cannot find our way soon—"

"Audrey."

"—then Captain Okoli will—"

"Audrey." The softer interjection came from Caspian.

"What is it!" she snapped, looking up from her pocket watch.

Caspian stepped forward, placed his hands on her shoulders, and gently swiveled them toward the city hall.

An enormous clock protruded from the white stones, two-thirds of the way up the wall. The clock resembled the silver pocket watch that now lay forgotten in Audrey's hands as a chandelier resembles a single candle flame. Audrey couldn't quite pick out the clock's hands, as though she were gazing at them through a mist or through tears, but she could detect erratic motions, as though the hands were jigging back and forth a little. The second hand mostly circumnavigated the clock face steadily but sometimes sped up, slowed down, or jumped backward.

Audrey gazed silently for a moment.

"On the other hand," she murmured, her eyes never leaving the clock face, "someone who knows the way to the docks is bound to show up and point us in the right direction. I have no doubt that we can afford to wait here another few minutes."

TIME RUNS only forward, according to the second law of thermodynamics. Wedgwood teapots smash but don't re-collect themselves; the ashes in a fireplace don't reconstitute kindling; we apologize but can't undo our betrayals. Clocks measure time, so they fall under the umbrella of thermodynamics—or, in the spirit of quantum steampunk, under the parasol of thermodynamics. You could regard quantum clocks as quantum thermal machines, but they merit their own chapter.

Autonomous clocks run by themselves, without the need for any winding or other control. You'd want an autonomous quantum clock if you were building an autonomous quantum machine. For instance, imagine a quantum drone navigating a molecular landscape alone. The drone would need to monitor its speed, as well as halt or unload cargo at specific times. The robot would have to carry a clock that wouldn't decohere the robot and that could operate without a power outlet. In another example, consider a quantum computer churning through a calculation without anyone typing commands into its console. An autonomous quantum clock would tell the computer when to apply the next logic gate.

Autonomous quantum clocks differ from the "atomic clocks" that you can buy in stores. The only stores from which you can buy autonomous quantum clocks exist in theorists' imaginations because we haven't built autonomous quantum clocks yet.

My grandparents bought an ordinary "atomic" clock and hung it in their kitchen. The timepiece was more accurately called a *radio-controlled clock*. Every day, the clock received radio signals from a Colorado outpost of the National Institute of Standards and Technology (NIST). The radio waves helped my grandparents' clock synchronize with a highly accurate clock operated by NIST.

NIST owns the world's most accurate quantum clock—or, at least, it did when this book was written.[1,2] The clock resides at the NIST's Boulder, Colorado branch—a sister to the Maryland branch to which I'll belong. The clock consists of an aluminum ion that interacts with a laser.*

Initially, the ion occupies a low rung on its energy ladder. Scientists shine a laser at the ion, which might absorb a photon. Suppose that the photon's energy isn't quite the difference between the energy of the ion on an upper rung and the energy of the ion on the lower rung. The photon carries too much energy or too little energy to be absorbed by the ion, so the ion might not jump. After shining the laser, the scientists measure whether the ion has jumped to its upper rung. If not, the scientists adjust the laser—they change the photons' energy—and try again. After many rounds of adjustment, the ion almost definitely jumps. The laser now emits photons of energy very close to the difference between the rungs' energies. We can calculate this difference using quantum theory. So, we know each laser photon's energy with high accuracy.

We can use our knowledge of the laser's energy to measure a second of time. Imagine shining the laser at a wall (figure 8.1). The laser light not only consists of photons, but also has wavelike properties. A wave crest hits the wall, and then another crest, and then another, till you shut off the laser. The time between crests, we can calculate from the laser's energy. We know the energy with

* We can say that an atomic clock contains an ion because an ion is like an atom, except it has a different number of electrons.

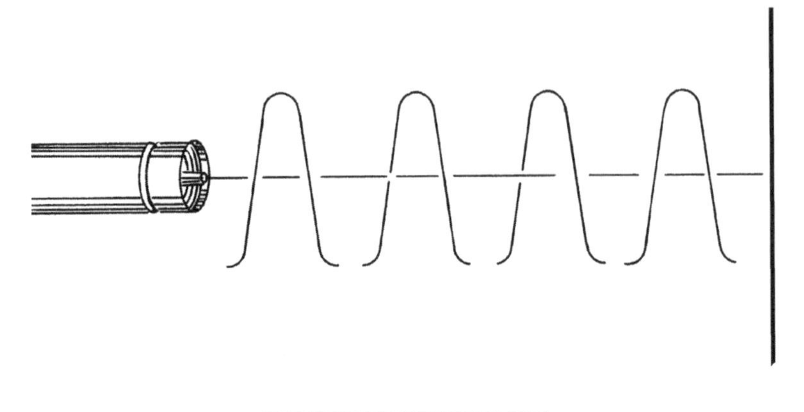

FIGURE 8.1

high accuracy, so we know the inter-crest time with high accuracy. That time is some fraction of a second.[3] To measure how long a second lasts, you wait until some number of crests hit the wall. You can measure a second of time with high accuracy, using an atomic clock. Such accuracy facilitates everything from the Global Positioning System (GPS), to the time-stamping of financial activities, to the management of the power grid. That's why I felt embarrassed when, more than a day into my first visit to Boulder, I hadn't switched my analog wristwatch to Mountain Time.

These quantum clocks enable us to measure time accurately, but they aren't autonomous: someone—an external controller—measures the atoms and adjusts the laser. What sort of a system could serve as an autonomous quantum clock?

Wolfgang Pauli pondered this question during the 1920s.[4–6] Pauli's exclusion principle explained why a two-fermion Szilard engine can't perform as much work as a classical engine or a bosonic engine (under certain assumptions). Pauli was an Austrian-American physicist whose interests extended from quantum physics to psychology and philosophy. He homed in on simple, yet fundamental concepts—and what is time, if not fundamental?

Pauli theorized about autonomous quantum clocks (I'll drop the *autonomous* from now on). The ideal quantum clock, he wrote, has a time observable. *Observable* is the physics name for a measurable property of a quantum system. We've discussed several observables—energy, position, momentum, and spin components—although I haven't called them observables till now. An ideal quantum clock could occupy a quantum state in which the time observable has a well-defined value.

In such a state, the energy wouldn't have a well-defined value, thanks to quantum uncertainty. According to the uncertainty principle, an electron with a well-defined position is in a superposition of all possible momenta. The more well-defined the electron's position, the less well-defined the momentum. Energy and time participate in a similar trade-off. So, a quantum system with a well-defined time would be in a superposition of all possible energies. This superposition would have an important property: Imagine preparing a system in this superposition and then measuring the energy. Your probability of obtaining one possible outcome equals your probability of obtaining any other possible outcome. That is, a system with a well-defined time is in a superposition spread evenly across all possible energies.

This fact reveals a trade-off between timekeeping and work extraction. Suppose that you have one quantum system, from which you want to extract work or which will keep time. A system performs work reliably if you can predict roughly how much work it'll provide. You can predict accurately if you know how much energy the system has—if the system has a well-defined energy. So, a well-defined energy often facilitates work extraction, while a poorly defined energy facilitates timekeeping. The ability to tell time trades off with the ability to provide work in quantum thermodynamics.

Pauli proved that no quantum system can have a time observable. If a system did, it could have an infinitely negative amount of energy. Having an infinitely negative amount of energy is impossible in our world. So, our world doesn't accommodate time observables—or ideal quantum clocks.

Timeless Timekeeping

So much for the autonomous quantum Rolex. But a quantum clock can approximate the ideal quantum clock, similarly to how a real engine can approximate a Carnot engine. Three colleagues of mine detailed the approximation, and I visited two—Jonathan Oppenheim and Mischa Woods—in London one spring. Jonathan is a professor of physics and astronomy at University College London. Mischa was a postdoc in a superposition between London and Delft, Holland. I'd spend all day working with them, and then walk to the British Museum in the evening. Although I'd arrive after the museum closed, I couldn't stay away. Like Audrey's parents, I've relished studying the ancient Near East and ancient Egypt—although the Stoqhardts have more expertise than I. The British Museum boasts a treasure trove of artifacts, which I intended to pore over during the weekend. What better way to ready oneself for millennia-old artifacts than by learning the latest science about time?

Mischa, Jonathan, and their collaborator Ralph Silva designed an approximation to the ideal time state, the superposition spread evenly across all energies.[7] My colleagues conceived of a different superposition of energies—a quantum state in which the energy is not maximally uncertain, but only highly uncertain.

Imagine measuring a quantum clock to learn the time, or controlling a quantum robot with the clock. You, or the robot, would interact with the clock. The interaction would disturb the clock, changing the clock's quantum state. The disturbance wouldn't interfere with timekeeping if the clock were ideal. But an imperfect clock would degrade, reducing our ability to distinguish instants.

You might as well gaze at a grandfather clock through increasingly blurry glasses: six o'clock will blend into 5:59 and 6:01, then into 5:58 and 6:02. Audrey noticed such blurriness when watching the quantum clock in Market Square at the beginning of this chapter.* Disturbances also hinder the clock's ability to initiate processes, such as logic gates in a computation, at desired times.

How well could Mischa, Ralph, and Jonathan's clock withstand such disturbances? Not too poorly, a Brit might say in an understatement. Imagine growing the clock—adding particles to it, although not so many particles that the clock loses its quantum nature. The bigger the clock, the greater its resilience. And giving a little gets you a lot: as the clock grows, its resilience grows exponentially.

Resilience also characterizes the British Museum's artifacts. I have a soft spot for *lamassu*, remnants of ancient Assyria. Ten-foot-tall statues of these winged bull-men guarded the entrances to palaces. Time has degraded the *lamassu*, but only a little: an observer can distinguish feathers in their wings and strands in their beards. Such artifacts are portrayed as having "withstood the flow of time," or "evaded the flow of time," or "resisted." Such portrayals don't appeal to me, although the *lamassu*'s longevity does. I prefer to regard them as surviving not because they clash with time, but because they harmonize with it in some way. From this perspective, *lamassu* lie only a few steps from the second law of thermodynamics and clocks.

On the other hand, the ancient Egyptians sculpted granite, when they could afford it. Gudea, king of the ancient city-state of Lagash, immortalized himself in diorite. Mischa, Jonathan, and I fashion ideas, which lack substance. Imagine playing, rather than rock-paper-scissors, granite-diorite-idea. The idea wouldn't stand a chance.

* Such a colossal quantum clock exists only in the imaginary novel that Audrey inhabits.

Or would it? Because an idea lacks substance, it can manifest in many forms. Plato's cave allegory has manifested as a story, as classroom lectures, on handwritten pages, on word processors and websites, in cartloads of novels, in the film *The Matrix*, and in one of the four most memorable advertisements I received from colleges as a high school junior. Plato's allegory—an idea—has survived since the fourth century BCE. King Ashurbanipal's lion-hunt reliefs, carved in alabaster, have survived for only 200–300 years longer.

The lion-hunt reliefs—and the *lamassu*—exude a grandness, a majesty as alluring as their longevity. The nature of time and the ideal clock have as much grandness, I believe. After leaving the British Museum's Assyrian gallery one Saturday, I boarded a train for Oxford. A quantum-thermodynamics conference was to take place there starting on Monday. I couldn't have asked for a more fitting follow-up to the museum.

Switching Gears

I accidentally discovered that an autonomous quantum clock appears to exist in the wild. The discovery owes its origin to David Limmer, a chemist at the University of California, Berkeley. David has a southwestern twang; the energy of a young scientist bent on changing his field; and, to me, the aspect of an academic older brother. He studies a molecule—found in nature and leveraged in technologies—depicted in figure 8.2.

The molecule contains two clusters of nuclei, represented by the spheres in the figure. The rods represent chemical bonds, or electrons shared by the clusters. You'll often find such a molecule in the "closed" configuration at the top of the figure. If you shine light on the molecule, one cluster may rotate around the other. The resulting "open" configuration appears at the bottom of the figure. The ability to switch configurations lends the structure one of its names, *molecular switch*.

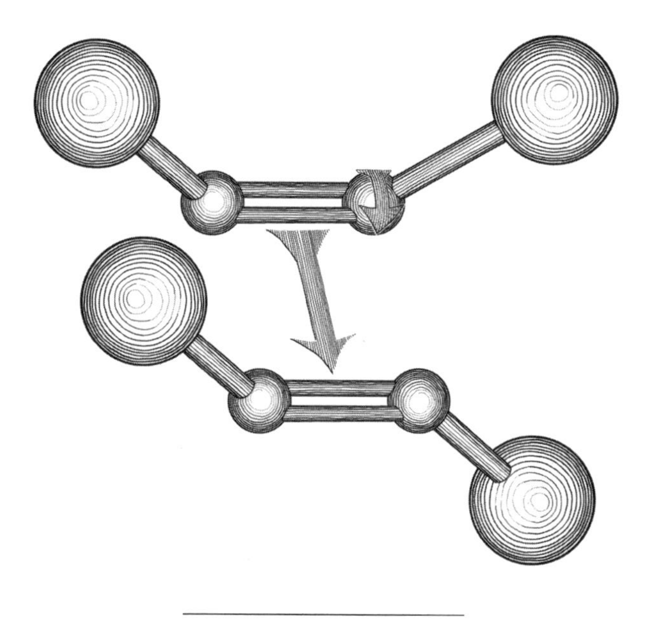

FIGURE 8.2

These switches exist in our retinas. When light enters our eyes, a molecular switch can absorb a photon. Changing configuration, the molecule hits a protein, as you might hit a bedside lamp when stretching after a night's sleep. The knock sets off a sequence of chemical reactions, which can end in the impression of sight. Therefore, these molecular switches matter.

David wanted to model the switch using quantum thermodynamics, for a reason I'll explain in chapter 11. He wrote down what he knew about the molecule; I wrote down mathematics from quantum thermodynamics; and we kneaded the two together.[8] The mathematics that represents David's molecule, I realized, contains mathematics that represents quantum clocks.

Think of the molecular switch as an autonomous machine that contains a quantum clock. The clock's hand consists of the rotating clump of nuclei. As a hand rotates down a clock face, so do the nuclei rotate downward. The right hand effectively points to 2 when the molecule occupies the upper configuration in figure 8.2. When the molecule occupies the bottom configuration, the hand effectively points to 4.

No human uses this clock to tell time. Rather, the clock tells the rest of the machine how to behave at each instant. The rest of the machine, in the molecular switch's case, consists of electrons. Nuclei account for most of the molecule's weight; electrons account for little. They flit about the landscape shaped by the atomic clumps. That landscape helps determine how the electrons behave. So, the electrons form the rest of the machine controlled by the nuclear clock. Earlier, we imagined a quantum clock that determines when a quantum computer performs some logic gate. You can replace *a quantum computer* with *the electrons* and replace *performs a logic gate* with *flit around a particular landscape*. So, a quantum clock determines when the electrons flit around a particular landscape.

A good clock's hands have well-defined positions: they report one time at each instant. Yet the hands also have well-defined momenta, traversing the clock's face at a constant speed. A quantum system can't have a well-defined position and a well-defined momentum simultaneously. So, how can a quantum system serve as a clock hand?

I learned an answer from the Londoner I visited between conversations with Mischa Woods and Jonathan Oppenheim: David Jennings is a quantum thermodynamicist who, at the time, worked at Imperial College London.* David reminded me that as a quantum system grows, it approaches classicality. Imagine starting with one quantum particle and adding other particles to it, or growing

* As University College London lies near the British Museum, Imperial College London neighbors the Victoria and Albert Museum and London's Museum of National History. Accordingly, I must confess to having arrived in the neighborhood hours before our conversation started.

the particles' masses. Repeat this process many, many, many times, and the system will wind up described by classical physics. A classical system can have a well-defined position and a well-defined momentum.

So, a quantum clock hand should consist of many quantum particles, or of massive quantum particles. They'll occupy the midlands between one quantum particle and a classical clock hand. Such a midland clock hand can occupy a quantum state in which it has a fairly well-defined position—if not a completely well-defined position—and a fairly well-defined momentum. Imagine preparing the clock hand in this quantum state, measuring the clock hand's momentum, and dividing the outcome by the clock hand's size. Repeat this process in each of many trials. Most of the resulting numbers will lie close to each other, so the hand has a fairly well-defined momentum. A similar story features the clock hand's position.

David Jennings's insight explains how David Limmer's molecule can serve as a quantum clock: The clock hand manifests as a clump of nuclei. The clump has a large mass relative to—for example—an electron. So, the clump can rotate fairly steadily and report times fairly accurately.

David Limmer studies these molecular switches partially because experimentalists can fabricate and control them easily. For instance, experimentalists can set the atomic clump moving—can "wind up" the clock—with lasers. All the other (autonomous) quantum clocks I know of live in our imaginations. But those clocks have been thriving: Mischa, Jonathan, Ralph, David Jennings, and many other theorists have been proving a flurry of mathematical results about them. Can molecular switches bridge quantum clocks from theory to experiment? Whether measured on a quantum clock or a classical one, time will tell.

CHAPTER 9

UNSTEADY AS SHE GOES

FLUCTUATION RELATIONS

Audrey gripped the ship's rail, shut her eyes, and tried to avoid remembering how many spoonfuls of marmalade she'd eaten at breakfast. Salty water kept spraying her face and dress, as though the sea were determined to reach out and befriend her. Audrey approved of cordiality generally, but she determined to decline this calling card. A robust friendship required sharing, her mother insisted, and Audrey preferred not to share her marmalade with the waves.

"All right there, miss?" The first mate, his face sunburned and wrinkled like an old peach, appeared at her elbow.

Audrey opened her eyes for an instant, shut them, and swallowed. She liked the first mate, and she usually liked peaches, but the thought of them did her stomach no favor now.

"Captain Okoli said to fetch anythin' y'need if you're unwell—and to ask you to come inside again. The sea en't rough to us as knows it, but 'tis different to a gentle lass such as yeself."

Audrey inhaled deeply through her nose and managed a "thank you" before the ship pitched again.

"Is it—is it always like this?" Audrey asked. Her jaw then clamped shut of her own accord, and she gave up on speaking more.

The first mate lifted his head to gaze at the green hills that were approaching. "Near Fluctuarian Bay?" he said. "Aye, miss, 'tis always like this."

THE SECOND LAW of thermodynamics, we heard from Sir Arthur Eddington, reigns as queen across the theories of physics. Could anyone improve upon it? Over the past three decades, we've learned that the answer is *yes*.

Fluctuation relations are equalities discovered in nonequilibrium statistical mechanics, a field that covers DNA strands, molecular motors, and the universe in its infancy. Fluctuation relations trump the second law in multiple ways. First, imagine a thermodynamic system out of equilibrium—say, our favorite classical gas in a box, with all the particles clumped together in one corner. The second law decrees that the gas's entropy will grow or remain constant. This decree encodes an inequality: the entropy at a later time must be at least the initial entropy. The inequality doesn't reveal how much the entropy will grow, as an equality would. So, an equality would provide more information, and fluctuation relations are equalities.

The alert reader will protest, "But the second law turns into an equality when the system reaches equilibrium, when the gas is spread across its box: the entropy has peaked, and it can't decrease, so it'll stay constant forever." Keenly observed, alert reader. But fluctuation relations are equalities even when the system is far from equilibrium—for instance, when the gas is roiled up because the box is dropped. Also, fluctuation relations enable us to predict more about single trials of an experiment than the second law does, as we'll see later.

Let us rephrase the virtues of fluctuation relations in steampunk terms. We'll draw upon the United Kingdom's National Archives, which has preserved a trove of Victorian-era ads—or adverts, as Audrey might say.[1] Examples include

"CALVERT'S CARBOLIC PRICKLY HEAT & BATH SOAP"

and

"THE CELEBRATED ACME WEED KILLER"
(with "NO SMELL").

A Victorian-spirited advert for fluctuation relations would read:

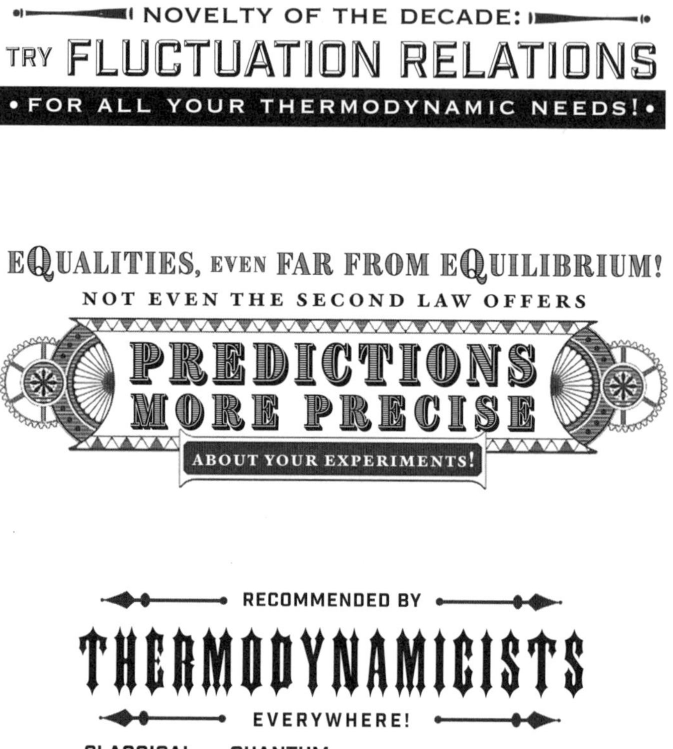

We'll start with classical fluctuation relations. Two of the most famous relations can be understood through a DNA molecule that's jolted out of equilibrium by being stretched quickly. Such experiments demonstrate that classical fluctuation relations work outside of theorists' imaginations. Theorists' imaginations, we've seen, have engendered many definitions for quantum work and heat. Work and heat feature in fluctuation relations; therefore, many quantum fluctuation relations have been formulated. Different ones describe different contexts, and some have withstood experimental tests. Whether quantum or classical, fluctuation relations demonstrate the growing role of experiments in quantum steampunk.

Fluctuations Are in Our DNA

Let's see a fluctuation relation in action. The top of figure 9.1 shows a length of DNA—two chains formed from complementary nucleotides. A loop, formed from more nucleotides, connects the top of one chain to the top of the other. The result resembles a hairpin, especially if you pry the two chains apart at the bottom.

The hairpin's thermodynamic behaviors obey the laws of classical physics; we have no need to describe them quantum mechanically. Nevertheless, the DNA's story merits telling here for three reasons. First, we can picture a DNA hairpin in our imaginations. Second, its story concretizes fluctuation relations, transforming them from mathematical abstractions into statements about our world. Third, experimentalists have performed the experiment that I'll describe.[2]

The DNA hairpin floats in water at room temperature. The hairpin begins at thermal equilibrium with the water and has some amount F_i of free energy. The subscript i stands for *initial*. F_i is the amount of energy you'd have to invest to create the DNA from nothing—to conjure the DNA's atoms from the void—and to warm the atoms to room temperature. You can think of such a conjuring as the pulling of a rabbit out of a hat.[3] F_i is also the energy that you could recover from the hairpin by annihilating it.

We can grab and manipulate the hairpin's two chains, using four tools. One is a handle attached to each chain's free end. The handle consists of more DNA. To the end of each handle, we attach a bead, tool number two. The bead measures about a micrometer across, like a bacterium. The third tool, a tiny pipette, applies suction to one bead, holding the bead steady on the pipette's tip.

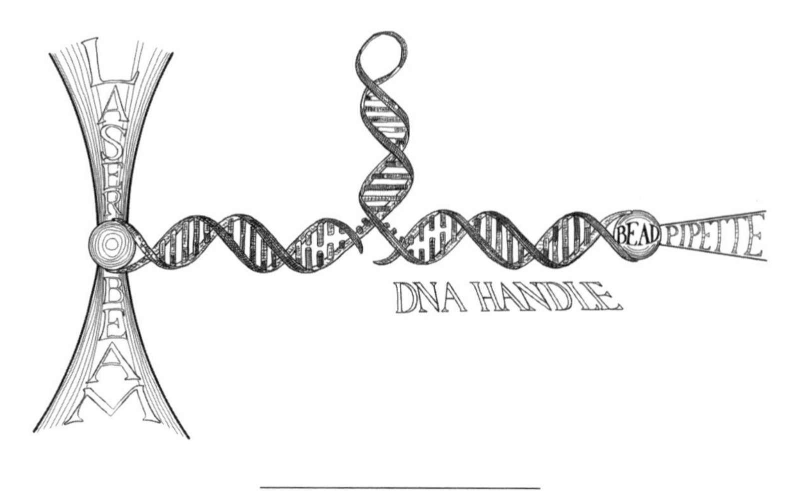

FIGURE 9.1

We grab and move the other bead using a laser, the fourth tool. The laser acts like tweezers akin to the ones with which my parents removed thorns from my fingers, when I touched our rosebushes as a child. Using the laser tweezers, we can pull the second bead through a predetermined distance. The pulling unzips the DNA hairpin, stretching its chains apart.

Stretching the hairpin costs work. The amount of work varies from trial to trial of our experiment: in one trial, a water molecule will kick the hairpin *here*, in *this* direction; in the next trial, a water molecule will kick the hairpin *there*, in *that* direction. So, the amount of work needed in a given trial is random. Imagine running many trials and stretching the hairpin through the same distance every time. We can measure the work required in every trial. The more times we have to pay a given amount of work, the more likely we are to pay that amount next time. So, from our measurements, we can infer the next trial's probability of costing a given amount of work.

The probabilities would obey a simple pattern in a special case:

at the end of the experiment, the DNA is unzipped. The pipette and laser hold the DNA's feet steady. If we held the feet there forever, the hairpin would come to thermal equilibrium with the water—to share the water's temperature. The DNA would end up with some amount F_f of free energy, wherein the subscript *f* stands for *final*. F_f is the energy we'd have to invest to pull the stretched hairpin out of a hat and warm the hairpin to room temperature. Imagine pulling a stretched hairpin out of a hat after annihilating a relaxed hairpin. The annihilation gives us an amount F_i of energy, which we can invest in creating the stretched hairpin from nothing. Since creating the stretched hairpin costs an amount F_f of energy, the whole process—the annihilation and creation—costs an amount $F_f - F_i$ of energy. Let's call this amount our *Boltzmann balance*, after the early thermodynamicist Ludwig Boltzmann.*

What relevance does the Boltzmann balance have to our DNA experiment? Imagine unzipping the DNA infinitely slowly, so that the hairpin always remains at equilibrium. The DNA wouldn't churn up the water, so we'd waste little energy. The randomness in our experiment would die down. We'd have to pay the same amount of work in every trial—and that amount would be the Boltzmann balance.

Chemists, biologists, and pharmacologists want to know how large this Boltzmann balance is. It governs how proteins change shape, how molecules bind together, how drugs diffuse across cell membranes, and more. But measuring the Boltzmann balance is tricky. For instance, you could measure the balance by stretching the DNA infinitely slowly, and measuring the work required, in one trial. But infinitely slow pulling would take forever. Fortunately, we can seek help from "FLUCTUATION RELATIONS FOR ALL YOUR THERMODYNAMIC NEEDS!"

* Scientists call the Boltzmann balance the *free-energy difference*, but I think that that's a horrid mouthful.

Fluctuation relations don't solve all thermodynamic needs—I'd never trust a Victorian advert—but they help resolve our problem: We can unzip the DNA quickly, jolting it out of equilibrium and jostling water molecules. The work spent to stretch the hairpin, we can measure. We'll repeat this pulling and measuring in each of many trials. From our measurements, we'll infer the next trial's probability of requiring *this* amount of work, or *this* amount, or *that* amount. We plug these probabilities into one side of an equation—a fluctuation relation—and, on the other side, out pops an estimate of the Boltzmann balance.

Christopher Jarzynski formulated this strategy—and the corresponding fluctuation relation—in 1997.[4] He's now a professor of chemistry at the University of Maryland. Apart from Chris, everyone calls the equation *Jarzynski's equality*. Chris is so humble, though, he calls it *the nonequilibrium fluctuation relation*.

Chris's equation offers a new way to measure the Boltzmann balance, but this way doesn't eliminate all difficulties. For instance, his method requires us to run many, many trials of the experiment. Fortunately, we can mitigate that challenge using information theory, as we'll see in chapter 10.

Jarzynski's equality doesn't merely help us estimate a Boltzmann balance—it doesn't only have technological applications. As a Victorian advertisement might say:

BUT WAIT: THERE'S MORE!

The fluctuation relation also illuminates fundamental physics.

Imagine unzipping the hairpin at your favorite speed in many trials. On average over trials, we spend an amount of work that's at least the Boltzmann balance. We can derive this result from the second law of thermodynamics, and we can derive it independently from Chris's equation. So, the fluctuation relation can sometimes replace the second law, if we care only about averages.

If we care about individual trials, the fluctuation relation reveals information absent from the second law. Imagine unzipping the hairpin at your favorite speed in one trial. You have some probability of spending little work—less than the expected amount, the Boltzmann balance. But that probability is small, and Jarzynski's equality stipulates precisely how small. So, the fluctuation relation tells us about individual trials, as the second law of thermodynamics doesn't. That's what the Victorian advert meant by **"NOT EVEN THE SECOND LAW OFFERS PREDICTIONS MORE PRECISE ABOUT YOUR EXPERIMENTS!"**

GOING INTO REVERSE

Jarzynski's equality is one fluctuation relation—one member of a class of equations. Some equations center on the entropy produced during an experiment, rather than on the work performed. Another work-based equation owes its origin to Gavin Crooks, a British-born physicist who lives in the United States.[5] His fluctuation relation has the fraught honor of being called *Crooks' theorem*, whose apostrophe scientists often misplace. "Crooks' theorem" appears in many papers, suggesting that the discoverer belongs in the penitentiary. Gavin bears the insult with patience and British wryness: The first message he ever sent me—a comment on a paper I'd cowritten about his theorem—read, "That's a most interesting paper you have on[line]. May I particularly commend you for putting the apostrophes in the right places."

Gavin compared Chris Jarzynski's experiment with its reverse. The reverse begins with an unzipped DNA hairpin stretched between a pipette and laser tweezers. The hairpin starts at thermal equilibrium with the water, having an amount F_f of free energy. The stretched hairpin resembles a stretched spring. As a spring would, the hairpin exerts a force on the tweezers that hold the hairpin taught. The experimentalist lets the DNA relax and zip up. The hairpin drags the tweezers toward the pipette, performing work on the tweezers. We can harvest the work extracted from the hairpin and use that work in another experiment.

How much work can we harvest? The amount varies from trial to trial, as with the forward (unzipping) experiment. We might expect to recoup as much work from reverse experiments as we spend in forward experiments. We do if we pull infinitely slowly: every reverse trial will provide the Boltzmann balance's worth of work. But we need to publish papers on our experiment before eternity dawns, so let's focus on reasonable speeds. We tend to recoup less work from reverse trials than we spend on forward trials. So, if we stretch the hairpin and then let it contract, we'll typically lose work, on balance. This net loss shares the spirit of the second law of thermodynamics, which has been synopsized as "You can't break even."

Gavin advanced beyond the slogan and the second law. He pinpointed how much more likely we are to lose a given amount of work, while unzipping the hairpin, than to win that amount while zipping. Let's compare the two probabilities at higher and higher stakes. As the amount of work grows, our probability of losing that work grows exponentially bigger than our probability of winning the work. We're far more likely to lose lots of work than to gain lots of work. Whereas most people say, "It's a dog's life," thermodynamicists can say, "It's a Crooks'-theorem life."

But Gavin's fluctuation relation doesn't only herald doom and gloom. It offers yet another means of estimating the Boltzmann balance coveted in chemistry, pharmacology, and biology. Consider unzipping the DNA at some realistic speed in each of many trials. From our measurements, we can estimate the next trial's probability of requiring *this* amount of work, or *that* amount, or *that* amount. Let's do the same for the rezipping experiment. One of our unzipping probabilities will equal one of our rezipping probabilities. The probabilities coincide when the amount of work equals the Boltzmann balance, according to Crooks' theorem. With help from

Crooks' theorem, we can estimate the Boltzmann balance in a new way from realizable experiments. Like Jarzynski's equality, Crooks' theorem offers practicality along with fundamental insight.

Imagine tailoring our Victorian advert for fluctuation relations so that it becomes an advert for Crooks' theorem. The flyer might excite itself into tachycardia, if a flyer could:

BUT WAIT: THERE'S MORE!!!

We can derive Jarzynski's equality from Crooks' theorem, just as we can derive a planet's trajectory from Einstein's equation for general relativity. So, Gavin's fluctuation relation precedes Chris's logically, although Chris's precedes Gavin's chronologically. In physics parlance, Crooks' theorem is *stronger* than Jarzynski's equality.

Any fluctuation relation—really, any equation—that describes far-from-equilibrium thermodynamics deserves a hurrah. Thermodynamicists originally focused on equilibrium, which is simple to describe because large-scale properties don't change in equilibrium. Discoveries came as thick as London smog. Later, thermodynamicists lifted their gazes to systems poked slightly out of equilibrium. For example, imagine bringing a weak magnet near a handful of iron shavings that form a thermodynamic system. The magnet nudges the shavings into changing their configuration. We can predict how quickly they change by using close-to-equilibrium thermodynamics.

What if the magnet is strong? It jolts the shavings, flinging them far from equilibrium, and most bets are off. Far-from-equilibrium thermodynamics resembles the Wild West: it's easy to get lost in; it's beholden to few laws; and it's filled with gunslingers difficult to tame. Rare is the equation that describes a wide swath of the far-from-equilibrium wilderness.

Yet fluctuation relations govern many systems far from equilibrium. We've seen that a DNA hairpin obeys fluctuation relations. So does RNA, which resembles DNA but consists of just one chain.[6] RNA folds up when left to its own devices, and you can unfold it with laser tweezers. Other single molecules obey fluctuation relations, too.[7] So do bacteria-sized beads[8] and tiny pendulums.[9] Aside from testing theoretical physics, these experiments redound upon other fields of science and medicine. For instance, understanding the tension in a protein can help us understand some genetic diseases.[10,11]

Classical mechanics describes the experiments that we've just surveyed. We can approach quantum physics by zooming in on single particles. A *quantum dot* is an artificial atom—in some cases, a little patch of space on a semiconductor surface. An electron is confined to the patch, by an electric field, similarly to how an electron in an atom is confined to remain near the nucleus. Why prefer artificial atoms to natural atoms? Advantages include manufacturability and our ability to tweak quantum dots' properties.

Imagine two artificial atoms sitting side by side. Each quantum dot contains electrons that can hop between the dots. We can influence the hopping by pushing the particles with an electric force: Electrons carry negative charges, so positive charges, such as protons, attract the electrons. Other negative charges repel the electrons. Imagine depositing positive charges rightward of the two dots and negative charges leftward (figure 9.2). The negative charges push the electrons rightward, and the positive charges pull the electrons rightward. (Experimentalists didn't actually deposit charges in this way; I'm simplifying the story. But the true setup pushed the electrons in the same way.)

Not only do those charges influence the electrons' motions. The electrons also receive heat from a bath that surrounds them, so they jiggle randomly. Instead of always hopping rightward, toward the positive charge, the electrons sometimes hop leftward. The pattern of hopping is random, and varies from trial to trial, because heat is random energy.

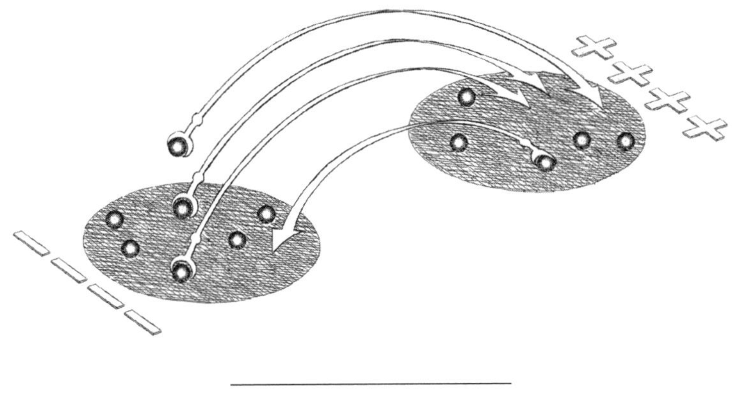

FIGURE 9.2

At the end of a trial, we can measure how many more electrons occupy the right-hand dot than the left-hand dot, using a certain meter. The imbalance implies how many electrons hopped rightward. Pulling and pushing each electron rightward cost the positive and negative charges work. So, we can measure how much work the charges have performed in a given trial. That amount of work satisfies fluctuation relations.[12,13] More-elaborate versions of this experiment have taken place in Finland, which I think a fitting locale for studying thermodynamics: Where can one better experience cold and appreciate heat?[14]

The quantum-dot experiments center on electrons, which seem quantum. Furthermore, the Finnish group detected hops by single electrons. Yet classical physics describes the experiments: The electrons share no entanglement and are in no other interesting superpositions. We can more fruitfully imagine the electrons as miniature kumquats than track the electrons' wavelike properties. Which is all well and good, as quantum steampunk encompasses small classical systems neglected by traditional thermodynamics, in addition to quantum systems. But fluctuation relations should extend to quantum physics. How can they?

Return to the Menagerie

The Victorian advert for fluctuation relations invited us to "TRY **CLASSICAL** AND **QUANTUM** FOR YOUR CONVENIENCE AND PLEASURE!" Developing quantum fluctuation relations can be less convenient than the advert suggests. To understand why, recall that a classical fluctuation relation governs a DNA hairpin that interacts with a bath while pulled by a force. The force imparts work, and the bath imparts heat. Like the hairpin, a quantum system can exchange work and heat simultaneously. But many conceptions of quantum work and heat exist; many species populate the menagerie of chapter 6. Different conceptions lead to different quantum fluctuation relations.

Some relations are easier to test experimentally, while some are abstract and mathematical. Some relations are true if the system slowly exchanges a little heat with the bath; some are true if the system exchanges a lot of heat quickly. Some fluctuation relations describe small classes of systems, and some describe large classes. Some relations describe high-energy particles, like those smashed together in experiments at CERN in Switzerland.[15–18] Another fluctuation relation describes the universe's expansion.[19] I developed a fluctuation-type relation that describes chaos, but we'll reserve that story for chapter 14.

Despite the diversity of quantum fluctuation relations, they share commonalities. You might have observed similar commonalities among a family's members. For instance, after living with a college roommate, you might have encountered a gaggle of strangers with her dark hair, pale skin, Canadian accent, and straight, white teeth and concluded, "These must be my roommate's relatives." In the family of fluctuation relations, all members are equations. Most interrelate quick, far-from-equilibrium processes with properties of equilibrium. They help us infer a useful number from measurable numbers. They involve the probability of performing a certain amount of work or generating a certain amount of entropy. These features enable us to recognize equations as fluctuation relations, despite their variety.

Will one quantum fluctuation relation rise to the top of the pile, like a prince who's poisoned and gored all the relatives who've fought for the throne? Will quantum thermodynamicists agree about how to define work and heat? I expect not. Call me a thermodynamic pluralist, but I see merit in multiple approaches. Which definitions and equations are useful depends on which system you're holding, how you poke it, and how you can measure it.

This plurality of equations contrasts with the oneness of many other scientific results. For example, imagine dragging a tea trolley across a Persian rug. Isaac Newton's second law of motion helps determine how much force you apply to accelerate the trolley by some amount. Not Newton's second law, version 1.A.iii, applicable only to Persian rugs—in contrast with version 1.A.iv, which governs Moroccan rugs. One law governs both situations, in contrast with quantum fluctuation relations. Perhaps, like Newton's second law, some principle will unify the quantum fluctuation relations, revealing them to be different sides of a multidimensional coin. Or perhaps quantum thermodynamics, in light of its variety, is richer than classical mechanics.

Following theorists, experimentalists have waded into the morass of quantum fluctuation relations. For example, Shuoming An and collaborators tested a relation based on the Yorkicockasheepapoo definition of quantum work and heat:[20] They separated heat exchanges from work exchanges in time. Furthermore, they measured the system's energy before and after each exchange.

In their experiment, an ion replaced the DNA hairpin. Yet the quantum experimentalists, like their classical counterparts, used lasers. The lasers trapped the ion as a well traps an apple that's fallen in. Trapping the ion involved cooling it to a low temperature, so that the ion wouldn't jiggle much. But the lasers and other equipment—like all equipment created in this mortal coil—suffered from imperfections. Such equipment operates when energy drives electrons around a circuit. A little of the energy escaped from the circuits, dissipating as heat. The experimentalists waited for the heat to warm up the ion a little, to act as the bath.

The experimentalists then measured the ion's energy, in accordance with the Yorkicockasheepapoo definition of quantum work and heat. Then, the experimentalists moved the laser beams. The lasers, trapping the ion, forced the ion to scoot. This force accomplished work on the ion. After the scooting, the experimentalists measured the ion's energy again. The ion's final energy minus its initial energy was the work performed on the ion by the lasers.

Experimentalists have used not only lasers and ions to check quantum fluctuation relations. They've also leveraged techniques used to photograph people's brains in MRI machines.[21] The MRI experiment, too, involves the Yorkicockasheepapoo definition of work and heat. But that definition hasn't dominated all the menagerie's visitors: One lab checked a fluctuation relation by weakly measuring superconducting qubits, proffering nectar to the hummingbird definition.[22] Superconducting qubits served also in another experiment;[23] atoms served in another;[24] and defects in diamond, in another.[25]

Fluctuation relations, quantum and classical, have withstood a wealth of experimental tests. This wealth demonstrates the partnership that welds theory to experiment across most of physics. The wealth also exemplifies the rise of experiments in quantum steampunk. For decades, quantum thermodynamics attracted mostly theorists. But experimentalists have learned to manipulate diverse quantum systems recently, to realize quantum computing and other technologies. Experimental techniques pioneered for technology are now informing science—especially quantum steampunk, as exemplified by fluctuation relations.

But Fluctuarian Bay, as the first mate in Audrey's novel called it, doesn't only connect theorists to experimentalists. Fluctuation relations also connect, themselves, to another realm on the map of quantum steampunk. To arrive there, we'll need to don our overcoats, disembark from our boat, and watch the horizon.

CHAPTER 10

Entropy, Energy, and a Tiny Possibility

ONE-SHOT THERMODYNAMICS

"That pickering-bus is your sole chance of reaching Singledon before the summer rains begin." Captain Okoli pointed into the sunrise.

The three travelers shaded their eyes to look. A vehicle was . . . Audrey struggled to find a verb that encapsulated the machine's action, other than *approaching*. The pickering-bus consisted of a green dome atop four spindly, jointed legs like a praying mantis's. The vehicle sprang from leg to leg, stalking toward the figures huddling on the filling station's rooftop.

"The pickering-bus shall dock directly below us," Captain Okoli continued, never taking his eyes off the vehicle. "Filling the tank takes nineteen minutes, during which time you shall drop three-and-a-half feet onto the deck." He pointed downward. "On the deck is a cushioned bench, with fifteen leather straps. You must tie those to your waists, legs, and arms, and not untie yourselves until arriving in Singledon."

For the first time, Captain Okoli turned to the three travelers. Audrey and Baxter quailed under the fierceness of his gaze, and even Caspian blinked twice.

"Remember," Captain Okoli continued, "you must *not* untie yourselves, on pain of a most uncomfortable death. When you arrive in Singledon, my sister shall lead you through a hatch unknown to the crew."

Audrey, Baxter, and Caspian turned their gazes back to the pickering-bus springing toward them like a fairground ride that no one wishes to brave.

"On pain of death?" said Baxter.

"A most uncomfortable death?" echoed Audrey.

"How do you know all those details about the bus?" asked Caspian.

Captain Okoli turned back toward the pickering-bus and watched it for a moment. A smile spread across his face, as though he were watching a kitten playing with a ribbon rather than a contraption that resembled a giant predatory insect.

"My sister and I invented the pickering-bus," said Captain Okoli.

"Oh," said Baxter.

"An excellent reason," added Audrey.

TRADITIONAL THERMODYNAMICS has a dirty secret that I haven't mentioned: the theory best describes infinitely large systems. That many-particle gas in a box we keep discussing? We should be thinking of it as containing infinitely many particles and filling an infinitely large volume.

When the gas is infinitely large, many measurable properties stick close to their average behaviors. To understand why, imagine yourself to be a newspaper reporter who covers cricket. For every team in the world, you track the mean number of runs scored per player, which I'll abbreviate as the team's *value*. You can average the value across the teams. Now, imagine that each cricket team grows to contain infinitely many players. Value will scarcely fluctuate across teams, as the vast number of players washes out discrepancies between teams.* Similarly, imagine preparing a gas in a box and measuring the mean energy per particle, or the *gas's value*. We prepare many such gases, measure their values, and average the values over the gases. Now, suppose that each gas grows infinitely large. Nearly every gas has a value close to the average. This infinite-particle regime, dominated by averages focused on in traditional thermodynamics, is called the *thermodynamic limit*.

* I'm imagining unrealistic levels of equality: that all teams enjoy the same amount of funding, that all players are of the same quality, etc.

The steam rising from a currant scone in the Stoqhardts' kitchen is outside the thermodynamic limit; the steam contains finitely many particles. The steam contains many particles, though—10^{24} of them. So, traditional thermodynamics describes the steam well for most purposes.

Still, traditional thermodynamics can dissatisfy us for three reasons. First, I've never seen an infinite number of cricket players—or particles, or anything else. So, describing our world with the thermodynamic limit rubs me the wrong way conceptually. Second, individuals sometimes concern us more than averages. The newspaper reporter draws in readers by cheering for an underdog and celebrating a superstar.

The rapper Eminem articulated the third reason in the song "Lose Yourself": Sometimes, "you only get one shot; do not miss your chance to blow. / This opportunity comes once in a lifetime, yo." For instance, Audrey, Baxter, and Caspian have only one opportunity to reach Singledon, and achieve their mission, before the summer rains arrive. The pickering-bus may halt at their dock, refill for nineteen minutes, and have fifteen leather straps on average. But the vehicle may be carrying an extra tank of fuel and not halt today, or an accident may have sawed off five leather straps. The travelers are wringing their hands not over the average trip, but over one shot.

Physicists and chemists have expanded thermodynamics to cover single shots. As throughout quantum steampunk, different people take different approaches. Some researchers focus on large deviations from the average—rare events that can make or break a mission. My experience lies in *one-shot thermodynamics*, which grew from a recent twist in information theory. To understand the twist, we'll return to the London life that the Stoqhardt siblings have briefly left.

When in London, we saw, Baxter should be standing guard at a cabal's headquarters every night. A probability distribution describes his likelihood of fulfilling his duty on any given night and his likelihood of having fallen asleep. Audrey records his performance in a journal. We imagined her monitoring him for 30 years, acquiring a string of about 11,000 letters. Audrey could compress this string into a few thousand bits—a few thousand units of information. How many bits would she need per letter in the original string? The number is the distribution's Shannon entropy—Audrey's uncertainty about what she'll find Baxter doing on the average night.

We can now recognize this story as approximating a thermodynamic-type limit. We imagined Audrey accruing 11,000 letters because 11,000 is much greater than one. Eleven thousand doesn't approximate infinity as well as 10^{24}, the number of steam molecules rising from a scone, does. But 11,000 letters lie far from one shot, thus resembling the thermodynamic limit. Accordingly, the answer to our question (into how few bits can Audrey compress her string?) involves an average. This average is the Shannon entropy (the average surprisal) because Audrey's record is classical. In the quantum version of Audrey's problem—wherein qubits replace bits—the von Neumann entropy replaces the Shannon entropy.

We aim to progress beyond averages, so we must progress beyond the Shannon and von Neumann entropies. This realization engendered a movement called *one-shot information theory*. One-shot information theorists ask questions such as, "Suppose that Audrey has recorded a string of three letters, and she wants to compress the string into the fewest possible bits. How many bits does she need?" If you answered "liver"—I mean, "an entropy"—you're correct. But the relevant entropy isn't one we've encountered; it was defined more recently.

Many of the entropies are, or build on, entropies defined by Alfréd Rényi in 1961.[1] We encountered Rényi as the Hungarian mathematician who quipped that mathematicians are machines for turning coffee into theorems. He could boast not only a quick wit, but also decades-long foresight. Today, one-shot information theorists dream up information-processing tasks, akin to data compression, and pinpoint the best efficiencies with which we can perform those tasks. Those efficiencies tend to depend on entropies, which tend to riff off Rényi's entropies. As the tech giant Apple has trademarked the slogan, "There's an app for that," information theorists can say, "There's an entropy for that."

One-shot information theorists invent entropies, catalog the entropies' properties (for example, entropies tend to grow, like the thermodynamic entropy), and tease out the relationships among entropies. These scientists remind me of phylogeneticists who not only discover species, but also concoct their own breeds. My friend Philippe Faist curates the phylogenetic tree of entropies, which resembles a tree less than it resembles a swampland.[2] A dizzying number of branches interconnect the species, and different types of branches denote different relationships. The most intrepid researchers have searched for the mama and papa entropies from which the rest descend, as evolutionary biologists seek the oldest cells.

These new-fangled entropies, we call *one-shot entropies*. They measure the best efficiencies with which we can perform single shots of information-processing tasks. Example tasks include data compression, the production of entanglement, and distinguishing which quantum state a system is in. From information-processing tasks, Szilard and Landauer taught, thermodynamic tasks are a hop, skip, and a jump away. So, one-shot entropies measure the best efficiencies with which we can perform thermodynamic tasks, such as work extraction. The study of these tasks, we can call *one-shot thermodynamics*.

One example builds on Landauer erasure. In chapter 5, we imagined that Audrey wishes to erase a qubit—to reset the quantum state to the quantum analog of a 0 bit. Audrey's qubit shares entanglement with a qubit of Baxter's. The siblings can erase Audrey's qubit while extracting work from the qubit pair and a heat bath. The siblings collect work, rather than spend work, by "burning" the entanglement.

Let's tweak the goal, acknowledging that the siblings can adopt any strategy they please. Some strategies are *deterministic*, or guaranteed to erase the qubit. Other strategies are *probabilistic*, having some probability p of failing to erase the qubit and a probability $1 - p$ of succeeding. Let Audrey choose her risk tolerance. If she carries smelling salts everywhere due to a liability to faint with fright, she'll choose a failure probability p close to zero. If drawn to gambling, fast horses, and skirts that expose her ankles, she'll choose a failure probability close to one.

Why accept a high probability of failing to accomplish your goal? Because risk trades off with reward. Choose a large p, and Audrey might slink away with a still-to-be-erased qubit. But, if she succeeds, she may hit the jackpot: the siblings may extract a great deal of work. A one-shot entropy governs the trade-off between risk and reward[3]—in this example and across one-shot thermodynamics.

{ BACK TO FLUCTUARIAN BAY }

One-shot entropies, we've seen, measure away-from-average behaviors and can describe small numbers of trials or particles. The close relationship between many particles and averages explains a feature of the experiments described in the previous chapter: Scientists have tested fluctuation relations with single molecules, single ions, and the like—with small systems that contain few particles. The systems' smallness enables us to check the extra predictions mentioned in the Victorian advert: **"NOT EVEN THE SECOND LAW OFFERS PREDICTIONS MORE PRECISE ABOUT YOUR EXPERIMENTS!"** Imagine trying to test fluctuation relations with many-particle systems, as by compressing the steam from a fresh-baked scone. Averages would dominate our observations. We couldn't detect evidence of the extra, more-precise predictions; we'd find support only for what fluctuation relations agree with the second law about: on average, compressing the steam costs an amount of work at least as great as a Boltzmann balance.

So, one-shot thermodynamics shares the spirit of fluctuation relations—a focus on small numbers. One-shot thermodynamics and fluctuation relations therefore resemble Audrey's acquaintances Miss Lillian Quincy and Mr. Raja from chapter 5: They *have* to meet.

This I concluded with collaborators at Oxford, early in my PhD. We determined to introduce fluctuation relations to one-shot thermodynamics, and introduce them we did.[4] A flurry of literature followed as colleagues dug more deeply into the intersection of fields. Whereas our results governed classical and quantum systems equally well, others delved more into nonclassical physics.

These colleagues and I are theorists; we proved mathematical statements and expounded upon their physical meanings. But our introduction of Miss Quincy and Mr. Raja bore fruit for experiments, thanks to Chris Jarzynski. I had the wonderful fortune to meet Chris months after releasing my first work about his equation. He was visiting Caltech to present the physics colloquium—the highlight of the department's week, the one talk attended by physicists from all subfields. A colloquium speaker, I've learned, follows a whirlwind of a schedule so packed that it rarely includes the time needed to walk from one meeting to the next. But Chris carved out a few hours to talk with me. As I explained about one-shot thermodynamics, he shared a problem that had niggled at him for years.

Jarzynski's equality enables us to estimate a number useful in chemistry, drug design, and biology: a Boltzmann balance, a difference between two free energies. We run many trials of a short, far-from-equilibrium experiment (as by pulling open a DNA hairpin), estimate the probability that the next trial will require an amount W of work, and plug the probabilities into a calculator. The calculator stirs the probabilities together with some other numbers, adds a teaspoon of algebra, and serves up an estimate of the Boltzmann balance.

This method is useful but suffers from a shortcoming: the smallest probabilities—the rarest outcomes observable in experiment—impact our estimate the most. To understand why, imagine that you're a government employee working for the British parliament during the mid-1800s. You're tasked with estimating the average wage earned in some textile factory. You stroll around the factory one Monday, count the children, count the men who perform this job or that, and count the women who perform that job or this. Then, you identify how much money a worker of each type earns. Say that you find mostly men in their 30s, lugging bales of cotton across the factory. The average wage could come close to their average wage—15 to 20 shillings per week.[5]

You're a diligent government employee, so you return to the textile factory on Tuesday and count the workers again. Again, the average wage lies between 15 and 20 shillings per week. The same routine unfolds on Wednesday. But, on Thursday, you find the factory's founder inside. He owns so many factories, and spends so many hours on paperwork in his office, he rarely visits. So, you have a minuscule probability of finding that any given worker, pulled from among the spinning jennies at random, earns as much as he does. But he earns so much money, your wage estimate skyrockets. If you hadn't returned to the factory on Thursday, your estimate of the average would have been far below the true average. The reason is, a rare event dominates the average wage.

Rare events also dominate the Boltzmann balance. How do they manifest in DNA-hairpin experiments? As trials in which unzipping the DNA costs only a tiny amount of work. Imagine unzipping the hairpin in trial after trial. Usually, the hairpin will be bucking like a bronco that we struggle to control, and the hairpin will have to knock many water molecules out of its way to unzip. We'll spend a pretty penny's worth of work. Once in a while, the hairpin will follow like a lamb when we beckon, and water molecules will nudge the hairpin along. We'll spend hardly any work.

Rarely can we get something for almost nothing. Even if we pull DNA hairpins till we want to pull our hair out, we'll likely never see the DNA docile as a lamb; we'll likely never witness the most important contributor to the Boltzmann balance. Therefore, we'll likely estimate the Boltzmann balance wrongly.

Worse, we can't easily tell how far our estimate lies from the truth. The assessment of errors forms a cornerstone of science. For example, particle physicists announced in 2012 that they'd discovered a particle, the Higgs boson, at a supercollider in Switzerland. The Higgs boson endows other particles with mass, and the discovery led to a Nobel Prize. But the experimentalists wouldn't declare the particle discovered until assessing their probability of having erred and until lowering that probability to 0.00003%.[6] Chemists using a Boltzmann balance likely wouldn't need such a strict standard. But still, they could be using an estimate different from the true Boltzmann balance. They'd need to know about how far the estimate lies from the truth.

How to measure that deviation is as straightforward as navigating London as a Californian unused to twisting, snaking streets. Why? The mathematical structure of Jarzynski's equality resembles a map that offers no guidance about how to translate inches on paper into miles on the ground: the map guides us toward our destination without revealing how close we've come. Jarzynski's equality helps us estimate the Boltzmann balance but doesn't reveal how close our estimate lies to the truth.

This shortcoming bothered Chris. So, he asked me if one-shot thermodynamics could help. How accurate is a free-energy estimate derived from his equation? If we need a certain accuracy, how many trials should we expect to have to run? Chris had estimated a number of trials,[7] but in a loosey-goosey manner that hadn't satisfied him. Could we reason about accuracy more precisely?

Indeed. We can recast Chris's question in the language of one-shot thermodynamics: Failing to witness a rare event is a failure in a thermodynamic task. We can choose our tolerance for failure—choose the greatest acceptable probability that a trial will fail. Similarly, we can choose what we mean by *rare event*—how little work we must pay to count a trial as rare. Our definition of *rare*, with our failure tolerance, dictates the number of trials we should expect to need to perform, to cap our probability of failing to witness a rare event (of failing to witness the factory owner's visit to the factory). That number of trials dictates our accuracy—the accuracy with which we estimate the Boltzmann balance.

A one-shot entropy, Chris and I showed, bounds the number of trials we should expect to need to perform.[8] So, one-shot information theory can guide experimentalists who use fluctuation relations. Our prediction governs, for example, a gas that's being compressed in a box (of course). If the gas contains only six particles, it's small enough to experience rare events—small enough that the compression requires anomalously little work in the occasional trial. The number of trials required to observe rare events obeyed our predictions.

Melding one-shot thermodynamics with fluctuation relations introduced me to the thrill of uniting disciplines. Audrey, on her journey across the parchment map, passes through city-states, villages, kingdoms, and principalities. They differ in language and dress, religion and cuisine. But the folks in one region sometimes face the same dilemmas as the folks in another region. The groups can benefit from comparing strategies, finds, and pitfalls. Alternatively, one country's citizens may invent tools that can end another kingdom's famine. I wouldn't ascribe such importance to my collaboration with Chris, but we did mitigate a fluctuation-relation problem by using one-shot thermodynamics.

One-shot thermodynamics trades goods with the next city-state that we'll visit on the map of quantum steampunk. If the city-states neighbored each other, their gates would stand open, and traffic would wend through, at all hours. But you can't reach the next city-state from one-shot thermodynamics by land. So, button up your cloak, and don't look down.

CHAPTER 11

RESOURCE THEORIES

A HA'PENNY OF A QUANTUM STATE

"Miss?"

Audrey's eyes fluttered open. Judging by the cold permeating her right cheek, she figured she'd fallen asleep against the curved wall of the dirigible's glass basket. She peeled her cheek off the glass, intending to rub warmth back into it, but her hand stopped halfway to her face when she glimpsed the airship's surroundings through the window. Audrey drew a sharp breath.

A network of islands hung in the air, tethered to each other and to the ground with chains of massive metal links. Woods filled one island, obsidian filled another, pillars of salt covered a third, and—Audrey gasped—nuggets upon nuggets of gold gleamed from the dirt on another. Audrey's vessel wasn't alone in navigating the islands. Other dirigibles floated past the islands, docked at them, and withdrew from the docks. About half the vessels bore flags, painted on their oblong surfaces, that apparently depicted the islands floating in the air. Nearly half the vessels bore flags recognizable from the city-state that the three travelers had just left. The rest of the dirigibles wore colors that she recognized from the previous week, or the week before, or colors that she didn't recognize but that—she suspected—she would grow acquainted with before her journey ended. The vessels, islands, and chains floated amidst wisps and curls of clouds, which fluttered like flags themselves.

"Miss."

Audrey tore her gaze away from the glass wall and turned to the freckled young conductor who was bobbing on his toes beside her seat. When he saw that she'd woken, his navy-and-gold cap bobbed farther toward her in a little bow (conductors appeared, Audrey noted, to wear navy-and-gold caps across the map and across transportation methods). He was holding the ticket that she'd placed above her seat, marking her territory and announcing her destination.

"Your stop is next, miss, and your friends farther down asked me to rouse you," he said, the navy-and-gold hat bobbing toward the end of the aisle. "We dock at the island in fifteen minutes."

WHAT DO YOU regard as a resource—what do you value? I value food, shelter, and warm clothing, which one can purchase with money—which therefore has value of its own. I appreciate time and sleep, which one can purchase to some extent. Family, friends, education, and fulfilling work rank highly on my list, as well. When I ask others what they value, they most often list food, time, and sleep. I've also heard "coffee" from several scientists—which I can believe, having had a colleague who'd arrive at the office when I was leaving around 7:30 p.m. and who'd leave when I was arriving around 8:30 a.m.

Beyond individuals, *organizations* value resources: Libraries need texts, and museums need specimens. Textile factories need workers; cotton, dye, and other materials; equipment; energy for powering the equipment; infrastructure; space for building the infrastructure; and money for building, for paying workers, and for purchasing materials. Beyond the organizations that you and I can build, *countries* value resources. Governments guard land, treasure precious metals, cultivate allies, and argue over intellectual property.

What distinguishes these resources as resources? Three properties, to my mind. First, resources are scarce: An oasis can water only so many of a desert's inhabitants. The Earth contains only so much gold near its surface. If animals excreted gold as we excrete urea, countries wouldn't have anchored their economies in gold.* Second, people value resources. Countries squabble over patches of ocean, companies invest in their workers' educations, and I protect my time from distractions. Why do we value resources?

* Also, you might think twice before gifting your sweetheart a gold bracelet—unless you want to break up with them gently—but that's another matter.

Because, third, we can use resources to accomplish tasks. Given time (and paper, a pencil, and web access), I can uncover secrets of the universe. Given a bootcamp on coding, workers can execute rote tasks more efficiently. Given coffee, mathematicians can prove theorems.

Resources drive politics, economies, and everyday decisions; and resources pervade our world. So, we'd benefit from a framework for analyzing resources. The framework would stipulate, for instance, how to quantify resources. Imagine spinning cotton into yarn with a wheel. How much yarn can we produce in a day? We can quantify the amount in pounds.

The framework should also enable us to compare resources: two pounds of yarn are worth more than one. We should figure out which resources can transform into others for free. If we own two pounds of yarn, we can throw half the yarn away; we can transform two pounds into one. Enacting such a transformation would be foolish, but the transformation is possible. In contrast, some resources can't transform into others for free. Given one pound of yarn, we can't turn it directly into two. Such a transformation is forbidden. However, we may be able to effect forbidden transformations by paying with another resource. Upon starting with one pound of yarn, we can end up with two by forfeiting money. In general, we should establish which transformations of resources are possible and which are impossible.

Some of my examples sound laughable: of course we can't transform one pound of yarn into two for free. But resources less straightforward than yarn—including quantum resources—exist. The question, "Can we turn this into that?" gives us pause when *this* and *that* are entangled quantum states.

Quantum information theorists have developed a toolkit for analyzing resources: the *resource-theory framework*. A resource theory is a model. You might have seen a model for a building—for a museum, or your country's capitol building, or a castle. Such a model is a simplified representation that fits on a tabletop. The miniature is simpler than the whole enchilada because it doesn't require as much material, or doors that open, or toilets that flush. The model captures essential features without complications. So does a physicist's model.

What does a "resource theory" model? Any situation with constraints that restrict which actions we can perform and which materials we can access. For instance, imagine that I'm a thermodynamic agent who expends work (such as to erase information) and extracts work (such as from heat engines). I'm in an environment at a fixed temperature—70 degrees Fahrenheit, according to the thermostat. I can easily access objects that are at 70 degrees: a tassel, a print of Robert van Vranken's steampunk painting, a brass key. Let's simplify the situation in our model: let's suppose that I can access any object I please, so long as it's at 70 degrees.

I could receive a non-70-degree item as a gift. For instance, I might have a friend who works in a lab down the hall. His lab could operate at 80 degrees, and he could gift me an 80-degree mince pie for the holidays. I'd receive a scarce object—a resource. In principle, I could press into service the 80-degree mince pie and a 70-degree mince pie as heat baths. I could extract work from them, using an engine, while the 70-degree mince pie warmed up.

In principle means *in theory*, which can mean *with extreme difficulty, so don't ask an experimentalist or an engineer to do this if you don't fancy a verbal hiding*. But Audrey finds her head in the clouds, at the beginning of this chapter, for a reason. We decided to distinguish what's possible from what's impossible, rather than what's practical from what's impractical. Let's countenance what's possible in principle and regard the mince pie as a thermodynamic resource.

We've established which objects I can't access easily. Which actions can't I undertake? We could list them all, starting with touching my tongue to my nose. But a thermodynamic agent who's trained as a contortionist might not obey all my constraints. Let's ignore them in our simple model; let's focus on the fundamental restrictions imposed on everyone by the laws of thermodynamics.

All thermodynamic agents obey the first law: the amount of energy in every closed, isolated system remains constant. I can shift energy from one system to another, such as by dropping a Wedgwood teapot onto a small paddle wheel. The teapot will lose gravitational potential energy; the paddle wheel will gain kinetic energy (energy of motion) and will spin; nearby air will gain energy from sound waves (*Kathunk!* goes the teapot as it hits the tabletop); and the tabletop will gain vibrational energy when the teapot lands. But the total energy of the teapot, paddle wheel, air, and tabletop can't change.

In our simple model, we assume that I can transfer energy in any way consistent with quantum theory. This assumption, like the assumption that I can summon a 70-degree first edition of the first Sherlock Holmes novel, is unrealistic. For example, physicists often change spins' energies by turning on a magnetic field—essentially, bringing magnets near the spins. But experimentalists can't change a magnetic field infinitely quickly. I can in the simple model, which emphasizes possibility over practicality.

In addition to sloshing energy from object to object, a thermodynamic agent can discard objects, or renounce control over them. For example, consider the teapot and paddle wheel. As a thermodynamic agent, I can drop the teapot, spin the paddle wheel, and so on. I can't recover the energy or information that dissipates into the air or the tabletop because I can't control the air molecules or the tabletop vibrations. The discarding of objects imbues our simple model with the spirit of the second law of thermodynamics, which stipulates that energy dissipates into uncontrollable parts of the world. We're not assuming that the second law governs our simple model; we're assuming only that I can easily access 70-degree systems and nothing else, that the universe obeys quantum theory, that I can slosh energy from object to object, and that I renounce control over some objects. From these assumptions, we can prove the second law, as well as stronger statements. But we'll arrive at those later; for now, we're setting the stage.

We've covered the three ingredients of a resource theory: First, easily accessible objects, we call *free systems*. These, such as objects at thermal equilibrium with the environment, aren't scarce. Everything that isn't free—such as an 80-degree mince pie, or an out-of-equilibrium system—is a *resource*, second. We may be able to use a resource to accomplish some task, such as performing work. The actions we can perform without consuming resources, we call *free operations*, third. A thermodynamic agent's free operations include plucking free objects from the environment, bringing them over to any resources we've been gifted, sloshing energy from one system to another, and discarding objects.

Together, a class of free systems, a class of resources, and a class of free operations characterize a resource theory. We've illustrated resource theories with a thermodynamic agent given a heat bath. But quantum information theorists have modeled cartloads of situations with resource theories.

The first resource theory ever formalized helped reveal the nature of entanglement. Nowadays, experimentalists can reliably create and control certain types of entanglement between certain types of particles. In one of many examples, experimentalists recently induced quantum dots to spit out maximally entangled photons on demand.[1] But decades ago physicists only dreamed of on-demand entanglement. And physicists are hulking, classical humans; we more easily destroy useful entanglement than cultivate it, even today. So, the first resource theory modeled two entanglement-challenged experimentalists who work in different labs. Call the experimentalists Audrey and Baxter. They can create no entanglement between their labs, as extreme versions of real experimentalists. They might receive a gift of shared entanglement—one electron, say, given to Audrey and an entangled partner electron given to Baxter. The experimentalists can attenuate this shared entanglement, as blundering classical folk do, by coupling their electron to particles in their lab. Each sibling can also manipulate their electron with magnets and can talk with their sibling via telephone. Unentangled states are free, or easy to create; and entanglement shared by the labs is a resource, or a rare boon.

Say that Caspian gifts the siblings an entangled state, giving one qubit to Audrey and its partner to Baxter. The siblings can only degrade and preserve the entanglement; they can't amplify it. How much does the entanglement degrade under a given action that they might take? Now, say that Caspian gifted the siblings one entangled state and the siblings want another. Can they convert his gift into the desired state? Can we rank all entangled states in a hierarchy, from most to least entangled? We can answer these questions by formulating the situation as a resource theory.[2]

Not that physicists thought of themselves as using a resource theory; the theory of resource-theories didn't exist then. Rather, physicists stumbled upon this approach to entanglement and found that the approach helped. So, they pinpointed what had helped: a simple model of free systems, resources, and free operations. They christened the model a *resource theory*, then spread the joy from entanglement manipulation to other settings. Quantum information theorists have now dreamed up resource theories to model quantum computation, communication channels, wavelike properties of quantum states, quantum logic gates, randomness (anathema to thermodynamicists but useful to cryptographers, who protect information from eavesdroppers), and more.[3] The resource-theory framework has billowed across quantum information theory like a mushroom cloud, leaving mathematical theorems instead of noxious gases in its wake.

The cloud spread to quantum thermodynamics early. The thermodynamic resource theory that we previewed when discussing mince pies—the model for an agent given a heat bath—emerged first. I've helped develop alternative models, but we'll visit those later. Fetch a sweater, if you share my tolerance for mildly cool weather, and let's explore life at 70 degrees.

Second Edition of the Second Law

To illustrate a resource theory for thermodynamics, let's recall the molecular switch of chapter 8—the molecule that can change configurations upon absorbing a photon. Say that the molecule begins in one quantum state—in the "closed" configuration, in a superposition of two energies. Into which quantum states can the molecule transition? More generally, if you're given one quantum state, which quantum states can you transform it into via free operations? We can answer such questions using a resource theory.

The second question—if you're given one quantum state, which states can you turn it into via free operations?—should remind us of one-shot thermodynamics: the latter concerns what a thermodynamic agent can achieve in one run of an experiment. As one-shot thermodynamics concerns single trials and resource theories concern single quantum states—both outside the purview of conventional thermodynamics—the two subfields overlap. I imagine, on the quantum-steampunk map, the town of one-shot thermodynamics on the ground and the federation of resource theories floating above.

The two questions generalize a question asked in conventional thermodynamics. In conventional thermodynamics, we consider loads of classical particles—for example, the particles carrying the scent of ginger and cinnamon from freshly baked gingerbread. The particles begin clumped together near the gingerbread's surface. We can imagine another arrangement of the particles, in which they're spread across the kitchen. Can the particles, we ask in conventional thermodynamics, transform spontaneously from clumped together to spread out? By *spontaneously*, I mean, without anyone's performing work on the system.

The second law of thermodynamics dictates how to answer this question: Check how much thermodynamic entropy the particles have when clumped together. Check how much thermodynamic entropy they have if spread apart. Check how the entropy would behave during the transformation. If and only if the entropy grows or remains constant, the transformation can happen spontaneously. The gingerbread steam's entropy would grow while the steam spread out, so the steam can spread out spontaneously.

Our conventional thermodynamic question has limited scope. First, the theory describes only a colossal collection of classical particles. Second, the initial and final states are equilibrium states. These assumptions govern only a handful of the transformations imaginable. For instance, our molecular switch consists of only one quantum particle, which can begin its switching while out of equilibrium. We drop these assumptions when shifting to the resource theory.

The resource theory can model any system whose energy is one of a few spaced-out numbers. The possible energies form rungs in a ladder, as for an atom. Quantum systems have energy ladders, and some classical systems' energies approximate a ladder. For example, imagine a rock on a short, vertical cliff. At most instants, the rock sits atop the cliff or at the bottom. So, usually, the rock has one of just two possible energies: the gravitational potential energy that comes from sitting high up or the gravitational potential energy that comes from lying close to the ground. So, the resource theory models quantum systems and approximately models some classical systems.

Using the resource theory, we can ask about any initial state—equilibrium, close to equilibrium, far from equilibrium, entangled or not, or otherwise wavelike or not. We can ask about any final state similarly. Let's morph the question, "Can this state transform into that state spontaneously?" into, "Can any free operation transform this state into that state?" In changing our approach, we've shifted gears from conventional thermodynamics to quantum steampunk.

How does a quantum steampunker answer that question? Multiple approaches exist, one rooted in mathematics developed by economists during the early twentieth century.[4] The economists aimed to measure how unevenly wealth was distributed across a population. For instance, imagine a population consisting of Audrey, Baxter, and Caspian. If each has five pounds sterling, then wealth is distributed uniformly. Now, suppose that Audrey has seven pounds, Baxter has five, and Caspian has five. Wealth is distributed less evenly. But wealth is distributed unevenly also if Audrey has six pounds, Baxter has four, and Caspian has five. Is one distribution more uneven than another?

The economists established a method for answering. Their questions turn out to differ from questions such as, "Is Audrey taller than Baxter?" We answer the tallness question by comparing one number that describes Audrey—her height—to one number that describes Baxter—his height. We compare many numbers in answering the question, "Is Audrey more athletic than Baxter?": we compare how quickly each sibling can run a mile, how far each can stretch, how much weight each can lift, and so on. The question, "Is this distribution more uneven than that distribution?" resembles the athleticism question, requiring the economists to compare many numbers. The economists identified which numbers to compare, analogous to identifying that the time needed to run a mile, the amount of weight liftable, etc. enables us to compare two people's athleticism.

Let's tweak the economists' question. Instead of asking, "Is this distribution farther from uniform than that?" let's ask, "Is this quantum state farther from equilibrium than that?" We can answer the question by purloining the economists' mathematics and generalizing it—by comparing a bunch of numbers characterizing *this* quantum state to a bunch of numbers characterizing *that* state.[5–8] If this state *is* farther from equilibrium than that state, this quantum state can transform into that via free operations (interacting the system with the fixed-temperature environment, sloshing energy between systems, and discarding a system), which tend to bring states closer to thermal equilibrium with their surroundings.

Purloining the economists' mathematics leads to an unexpected delight of using thermodynamic resource theories, like a chocolate chip in what you thought was a raisin scone: economic images crop up amid the thermodynamics, creating charmingly absurd juxtapositions. Examples include the image of Robin Hood, the medieval English outlaw who robbed from the rich and gave to the poor, striding through a quantum laboratory.

Imagine that Baxter owns all the wealth in England. Along comes Robin Hood, riding up to Baxter in the countryside one morning. Robin Hood steals a pound sterling from Baxter and gifts the pound to Audrey. Economists call this redistribution of wealth a *Robin Hood transfer*.

If a Robin Hood transfer turns one distribution of wealth into another, the final distribution is more uniform—more evenly spread across the population—than the initial. As Robin Hood transfers feature in economics, they feature in quantum steampunk.[9] In quantum steampunk, a Robin Hood transfer moves not a pound sterling, but probability: Say that Audrey has an electron far from any magnetic field. She can measure whether the electron's spin points upward. Her detector has a 75% probability of reporting "yes" and a 25% probability of reporting "no." As a thermodynamic agent, Audrey can manipulate her electron. She can rob the "yes" outcome of some probability, which she transfers to the "no" outcome. This thermodynamic Robin Hood transfer turns the outcome probabilities into 50% and 50%. The probabilities are now spread evenly, and the state is in thermal equilibrium.

As Robin Hood transfers bring societies closer to equity, they bring quantum states closer to thermal equilibrium. The operations performable by a thermodynamic agent tend to bring quantum states closer to equilibrium. So, economic mathematics can help answer the question, "Can any free operation transform this state into that state?" Answering the question fully requires us to augment the economic mathematics, which can't describe quantum systems' wavelike properties.[8]

I hinted that multiple approaches can answer our question. If you guessed that one approach involves liver—I mean, entropies—you receive a gold star. This approach requires us to tweak the question. Suppose that a molecule would undergo the transformation in question. That molecule transforms in the presence of other systems—for example, laboratory equipment: magnets used to manipulate the molecule, the clock used to time the manipulation, and the tabletop on which the atom sits. This equipment may degrade slightly as it's used. We should incorporate this degradation into our question, which becomes, "Can any free operation transform this state into that state while the equipment used degrades a little?"

We can answer this question using liver—that is, entropies. The use of entropies is fitting: Recall our conventional-thermodynamics question, "Can this equilibrium state of a large, classical system turn into that equilibrium state spontaneously?" We answered by checking whether the system's thermodynamic entropy would decline or stay constant. Similarly, we answer the resource-theory question by checking whether many entropies decline or remain constant.[10] These entropies build on those defined by Alfréd Rényi, one of our favorite Hungarian mathematicians.

I should advertise a disclaimer: these entropies sometimes flag when one quantum state can't transform into another, by shrinking during the prospective transformation. In other cases, one state can't transform into another, but we can't tell from the entropies, which rise or fail to change: quantum states can have certain wavelike properties to which the entropies are blind. So, we can never judge, from these entropies, that one state can transform into another. Still, the entropies can flag many transformations as impossible.

We've now identified two strategies for checking whether one quantum state can transform into another thermodynamically. One strategy involves mathematics from economics, and the other strategy involves entropies. Each strategy requires us to check whether a family of inequalities are true. Why a family? We had to check only one inequality in conventional thermodynamics—that the thermodynamic entropy doesn't shrink, in accordance with the second law. The answer is we can learn more from quantum thermodynamics—about entangled states, other nonequilibrium states, and small systems—than from conventional thermodynamics. Learning more costs more effort, as acing a class costs more effort than earning a C.

So, the resource theory implies predictions that conventional thermodynamics doesn't. Is the resource theory, then, unconnected to conventional thermodynamics? No; although we used the resource theory to evade averages and large objects, the resource theory *can* model averages and large objects. For instance, consider the entropies built from Rényi's. These entropies form a family with many members, outside the context of averages and large objects. In that context, though, the family members remind me of Edgar Allan Poe's poem "The Raven." The speaker is a young man who's startled, late one night, by a tapping sound. The tapping exacerbates his nerves, which are on edge due to the death of his love: "Deep into that darkness peering, long I stood there wondering, fearing, / Doubting, dreaming dreams no mortal ever dared to dream before." The speaker realizes that the tapping comes from the window, whose shutter he throws open. His wonders, fears, doubts, and dreams collapse onto a bird's form as a raven steps inside. So do the many entropies collapse onto one entropy as the system under consideration grows infinitely large. We could say, instead, that the entropies come to equal each other, but I'd rather picture "The Raven." So, the many inequalities collapse to become one inequality, which resembles the second law of thermodynamics.

So, the resource theory reinforces conventional thermodynamics while uncovering new ground. This status has endowed the family of inequalities with the title "second laws" of thermodynamics. The inequalities descended from economics deserve the title, as do the inequalities descended from Rényi's entropies. We could call fluctuation relations "new second laws," too. Some days, I feel as though everyone and their grandmother has derived their own second law. Even I've established some, with a collaborator.[11,12] Thermodynamics needs sharpening, and the resource-theory framework serves as a whetstone.

Endless Forms Most Beautiful and Most Wonderful

Resource theories sit atop a silver-tasseled pillow in my heart. Featuring entropies, and born from the desire to manipulate entanglement, they belong in quantum information theory. When featuring heat baths and energy conservation, they model thermodynamics. Resource theories have unveiled fundamental limitations on transformations, work performance, and work expenditure. They've extended the second law of thermodynamics, revealed how heat degrades quantum wavelike properties, and expanded the purview of one-shot thermodynamics.

But a soft spot mustn't blind us to faults, especially in science. Resource theories are unrealistic. First, thermodynamic agents can't conjure arbitrary 70-degree objects. Second, becoming able to slosh energy from object to object in every way permitted by quantum theory would cost years, and millions of dollars, of research and development. Third, the work extractable from a qubit wouldn't outweigh the work invested in the extraction, as discussed at the end of chapter 7: Cooling the qubit, until it behaves non-classically, requires energy drawn from an outlet in the wall. The outlet supplies more energy than the qubit supplies as a quantum engine. Fourth, resource theorists use abstract language, speaking of *systems* and *qubits*. Experimentalists speak concretely—of atoms, circuitry, lasers, pulse sequences (which stipulate when and how magnetic fields are turned on and off), and semiconductors (materials in, for example, your mobile phone).

Again, not for nothing does Audrey find her head in the clouds at the start of this chapter. I had my head in the clouds, too, having embarked on my quantum-steampunk journey in the federation of resource theories. I didn't regret my research; and I felt proud of my collaborators' and my contributions to pure, fundamental theory. But I felt compelled to ground my work. So, I turned around and, like a ceramicist throwing a vase back on the wheel, criticized my science.

Stage one involved generalizing, with a collaborator, the conventional thermodynamic resource theory. The resource theory that we've discussed models systems that exchange only energy. But thermodynamic systems exchange oodles of things: energy, particles of various types, electric charge, volume (if the systems are gases separated by a sliding partition, so that one gas expands by squishing the other), and more. A system can exchange different things by interacting differently with different environments. Such environments manifest in kitchens, in batteries, near magnets, in the Earth's gravitational field, in clumps of atoms cooled to near their lowest energy-ladder rung, and elsewhere. Charles Darwin wrote in *The Origin of Species* that biology encompasses "endless forms most beautiful and most wonderful." I find such a variety in thermodynamics.

To model this variety, resource theories needed generalizing.

I initiated the generalization with a collaborator, in graduate school.[11,12] We established a family of resource theories, which model thermodynamic exchanges of energy and/or particles and/or electric charge and/or nearly whatever else you please. We proved that further "second laws" govern which states can transform into which and which states can't transform into which.

Generalizing the resource theory opened up, if not "endless forms," then many forms—many thermodynamic systems and interactions—to being modeled in quantum steampunk. The more systems we can model, the greater our chance of finding some experimental platform that can test our results, or some part of nature that we could understand better through resource theories.

Not that our generalization has led to an experiment or to a better understanding of physics in the wild. Nor does our generalization model all the properties, listed in chapter 2, of quantum systems. But other steampunkers added those quantum properties to the model.[8] Meanwhile, the generalization contributed to the founding of another kingdom in quantum steampunk, which we'll see in chapter 12.

Also, the generalization expanded fundamental physics. We've seen that Landauer's principle stipulates that erasing information costs work. Physicists had devised a means of paying, rather than with work, with angular momentum.[13] Similarly, you can pay for a cream tea with cash or with a Visa card. We extended the payment methods to Mastercard, American Express, travelers' checks, Treasury bonds, stocks, and bartering, within the wallet of resource theories. Finally, the generalization stoked my angst about the abstractness of resource theories.

The angst had begun simmering during one of the most uncomfortable moments of my career. (I've notched many more-uncomfortable moments since then.) I'd just finished presenting a seminar about resource theories for thermodynamics at a university. My "Thanks for your attention!" slide remained onscreen, and the question-and-answer session had begun. My host—whom I'd regarded as a friend till that moment—pointed to a listener and said, "This is your arch-nemesis."

The audience consisted of physicists who studied quantum materials constructed in labs. They had their feet on the ground, unlike Audrey at the beginning of this chapter. My host labeled an experimentalist as my arch-nemesis.

"What implications does your theory have for his lab?" my host asked. "Does it have any? Why should he care?"

I could have answered better, but I had too little experience, having barely begun pursuing my PhD. I apologized that resource theories had sprung from the rarefied mathematical air of quantum information theory. I recalled the baby steps with which science sometimes migrates from theory to experiment. The crowd looked unconvinced, but I scored one point: the experimentalist was not my arch-nemesis.

"My new friend," I insisted.

Facing that experimentalist—as well as that embarrassment—made me strive to connect the resource theories in the clouds with the reality on Earth.[14] I started asking myself and others: Which pieces of resource theories merit testing with experiments? Which quantum states could an experimentalist prepare to confirm those pieces? What hope have we of testing experimentally whether certain transformations are impossible?

Also, how can theorists improve resource theories to render them more realistic? Can we bridge resource theories to experiments via another city-state on the map of quantum steampunk? For instance, experimentalists have tested fluctuation relations, which overlap with one-shot thermodynamics, which overlaps with resource theories.[15,16] Perhaps we could trace this chain from resource theories to experiments.

To my delight, other scientists took up the challenge. Some colleagues proposed experiments designed to check resource-theory results. Some modeled concrete physical systems in resource theories, such as atoms interacting with light. Some colleagues bridged resource theories to experiments via fluctuation relations.

Still, I couldn't shake my dissatisfaction with my research. I kept recalling the experimentalist who'd refused to turn his atom experiment into a quantum thermal machine. Extracting the greatest possible amount of work from a quantum engine wouldn't benefit him. Sure, he'd be checking our predictions. But we'd derived our predictions from quantum theory, which has withstood a century of tests. Another, equivalent test of quantum theory wouldn't advance humanity's knowledge. Besides, the greatest possible amount of work extractable wouldn't satisfy a flea. He could draw loads more work from a wall socket.

We encountered this experimentalist in chapter 7, when discussing quantum thermal machines. But I should confess, he was refusing to take up my challenge of bridging resource theories to experiments. I failed to answer my own challenge. Fellow resource theorists approved of the challenge, but other quantum physicists wouldn't give us resource theorists the time of day. Some smiled upon our proofs of fundamental limitations on thermodynamic transformations, as fundamentals interest most physicists. But other quantum physicists had no use for our resource theories; they refused to invest in the experiments we proposed. They wouldn't invest, I realized, until we showed that resource theories could accomplish something outside our suburb of physics. We'd have to apply resource theories to solve problems dear to other scientists' hearts. The perfect problem arrived shortly before I finished my PhD—not from another suburb of physics but from a whole other continent of science.

{ Crossing Borders }

We met David Limmer briefly in chapter 8, when discussing quantum clocks. But David, a theoretical chemist at the University of

California, Berkeley, merits more of an introduction. Meeting him in the last year of graduate school, while I was planning my postdoctoral years, returned me mentally to elementary school. There, I'd looked up to two students who were three grades above mine. They'd represented our school in science fairs, participated in speech competitions, and enrolled in rigorous high school programs. I look up to David as I looked up to those two students. He'd completed his postdoctoral stint two years earlier and was building his research group. He studies statistical mechanics far from equilibrium, using information theory and other mathematical tools. Although a theorist ardent about mathematics, he partners with experimentalists. He keeps an eye on topics as far afield as black holes. And he's even three years older than I, like those kids in elementary school.

Keeping an eye on topics far afield had brought resource theories to David's attention. He asked me whether we could use a resource theory to answer a question he'd posed about chemistry. The question concerns the molecular switch introduced in chapter 8 and depicted in figure 8.2. The molecular switch is called a *photoisomer*. Photoisomers appear across nature and technologies: we have them in our eyes, as mentioned earlier, and experimentalists have used the switches to improve the storage of solar fuel. The switch has two collections of bonded-together atoms, both attached to an axis.

Your average-Joe molecular switch spends much of its life in equilibrium, exchanging heat with room-temperature surroundings. The molecule has the shape at the top of figure 11.1, called the *cis* configuration. Imagine shining a laser or sunlight on the switch. The molecule can absorb a photon, gaining energy. The energized switch has the opportunity to, well, switch: one group of atoms can rotate downward. The molecule will come to occupy its *trans* configuration.

The molecule now has more energy than it had while in

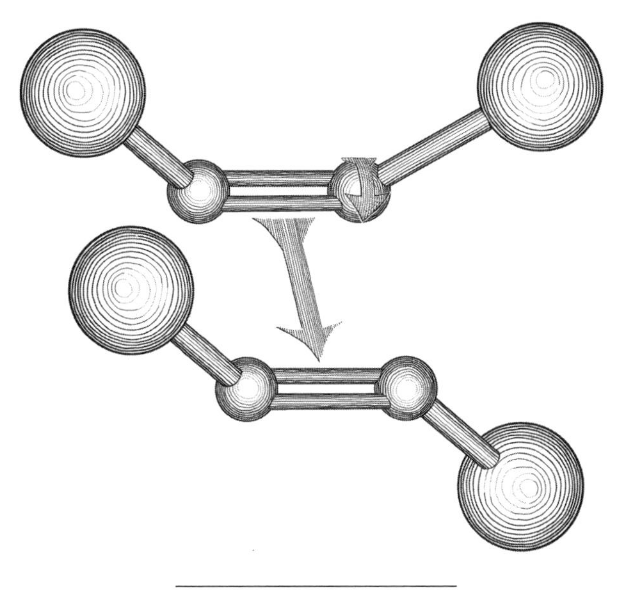

FIGURE 11.1

equilibrium, albeit less energy than it had right after absorbing the photon. The molecule can remain in this condition for a decent amount of time; that is, the molecule can store the photon's energy. For that reason, experimentalists at Harvard and MIT attached molecular switches to nanoscale tubes designed to capture and store solar fuel.[17]

Upon absorbing a photon, a molecular switch is well-poised to switch but might not end in the desirable rotated state. What probability does the molecule have of switching? This question has resisted a simple general resolution because molecular switches prove difficult to model: they're small, quantum, and far from equilibrium. We could build a model for the switch that incorporates every detail we know. But, the more detailed a model, the less general the model—the fewer situations the model can represent. We might prefer a simple model that represents many situations. So, David wanted to derive a simple, general bound on the molecule's probability of switching. Thermodynamicists specialize in such bounds. For instance, the second law of thermodynamics is a bound general enough to govern systems from quantum engines to the Earth.

David had a hunch that one could bound the switching probability using a resource theory. After all, resource theories are simple models. The first thermodynamic resource theory described in this chapter involves few assumptions: quantum theory, the law of conservation of energy, and the existence of a large heat bath.

David's idea lit me up like the photons impinging on a nanoscale solar-fuel device. He taught me about photoisomers, I taught him about resource theories, and we derived a bound.[18] First, we modeled the molecule in the resource theory (discovering that the molecule contains a quantum clock, as explained in chapter 8). Then, we chose mathematical theorems proved about resource theories and applied the theorems to the photoisomer. One theorem implied that quantum wavelike properties of the molecule can't help it switch. Physicists (and chemists) love discovering that quantum properties promote useful behaviors, so the realization disappointed me. But such is science, and we achieved our goal: we bounded the switching probability by applying "second laws" of thermodynamics to the molecule. The bound worked in computer simulations of the molecule. Furthermore, the bound lies close to the true switching probability if the molecule is unlikely to absorb a photon, as in the Harvard-MIT experiment.

To David's and my knowledge, our work is the first to answer a question developed outside quantum information theory, using a resource theory. Since we reached our goal, resource theories have been applied to condensed matter. Other physicists and I are applying them to black-hole physics. Quantum gravity inspired this brand of black-hole physics, so my head remains partially in the clouds—if not in outer space. But part of me mimics the merchant vessels glimpsed by Audrey through the dirigible's glass:

shuttling back and forth between the clouds and the Earth, working to ground one realm of quantum steampunk and lift others. Thinking in terms of resources—cups of coffee, pounds of yarn, a ha'penny's worth of entanglement or of distance from equilibrium—frees us to fly.

CHAPTER 12

The Unseen Kingdom

WHEN QUANTUM OBSERVABLES DON'T COOPERATE

The tent's burlap flap rustled, and an object scraped across the sand. Before Audrey, Baxter, and Caspian could register what had happened, the scraping ended, and footsteps pounded away into the night. Baxter reached the tent flap first, and he stooped to pick up the package that had been thrust inside.

"We have received a letter," he announced, untying the string tied around the folded brown paper. The three travelers returned to the little table where their lamp flickered and smoked. Baxter hunched over the paper as Caspian bent over his left shoulder and Audrey pressed against his right.

"Runes," said Caspian, who could see more than Audrey. "Give the letter to your sister, Baxter; she will have the best chance at translating it."

Baxter obeyed, and the huddle reorganized itself around Audrey. She held the paper up to her face, squinted, and ran a finger over the symbols.

"'Tis difficult to see," she said. "The writing is small at the end, but I believe I can nearly make it out." Her tongue poked between her lips as she concentrated. "To find the . . . unwitnessed . . . unknown? No, 'tis more like 'unwitnessed' than 'unknown.' Maybe—ah, I believe the word is 'unseen.' Very well, then; to find the unseen kingdom, walk—no, come—come along after—follow. Yes, follow the . . . arrow. Pointing. Sign? Ah, the signpost! To find the unseen kingdom, follow the signpost. Then comes a smudge that . . ."

Audrey brought the paper closer to her face, and the others bent closer over her shoulders. Suddenly, she flung her head back, nearly knocking skulls with both her companions.

"Baxter," she said, holding out a hand, "give me the Collection, please."

Baxter had fashioned the Collection himself, and he always kept it in a pocket. Since Baxter was Baxter, the pocket might be in the jacket he'd worn last week, but the Collection presented itself after he'd patted only three possible hiding places in today's outfit. One might be inclined to call the Collection a Swiss army knife; it contained a penknife, a corkscrew, a screwdriver, a magnifying glass, a ruler, a compass, and scissors. But Baxter had added a laser, a quantum sensor, an antenna, and a refrigerator large enough to house two qubits on a chip. He passed the Collection to Audrey, who teased out the magnifying glass and held it up to the page for a moment.

"My," she murmured. "This lies in your department, brother."

Audrey handed the paper and the Collection to Baxter, who brought the magnifying glass to one eye and then whistled.

"I have never seen a smaller refrigerator," he said. "I should like to meet the engineer who built it."

"What are we intended to do with a minuscule refrigerator affixed to the end of a mysterious letter?" asked Audrey. Silence fell as all three travelers ruminated over the delivery.

"Arrow," murmured Caspian. "Signpost."

"Eh?" said Baxter.

"Let me see the Collection," said Caspian.

He sorted through the tools that Audrey handed him until finding the quantum sensor, a defect-containing diamond shielded from its surroundings.

"The letter could have fallen into anyone's hands," said Caspian, "and the writer wanted for the signpost to be visible only to us. The signpost is likely small, judging by how the writing shrinks, and can be represented as an arrow. The best way to detect a tiny arrow is with a tiny sensor; and no one would be able to detect a tiny arrow, far from a laboratory, in the middle of the desert—apart from our Baxter."

Baxter didn't bother hiding the grin that crept from one ear to the other until Caspian added, "And, of course, apart from Ewart." But a needle was shifting across a dial on the sensor, which Caspian held up to the light after it froze.

"Down," he announced, before turning to Audrey. "Audrey, what lies below us?"

Audrey had read all about the region as soon as her parents had planned a dig there.

"Workers' huts from the fourteenth century BCE," she said.

"What lies below the huts?"

Audrey shook her head. "Excavations have focused on the palaces built by the workers, so no one ever bothered with the huts." She looked hard at Caspian. "Do you think that the lost kingdom lies below the huts?"

He shrugged.

"No one has seen the lost kingdom in millennia. It must be where nobody has bothered looking."

ONE OF MY favorite quantum-steampunk mysteries lay buried beneath the mundane—beneath "workers' huts"—for decades, if not for the millennia in Audrey's story. The "huts" manifested as a story told as often, in thermodynamics, as a folktale beside a hearth. Thermodynamicists often envision a small system interacting with a big bath. For instance, imagine trapping, in a bottle, the steam that rises from a freshly baked currant bun. Separate a handful of the steam particles from the rest with a piston. The rest can serve as a heat bath.

The handful of particles and the rest exchange stuff. The stuff might be heat, if the piston transmits energy; or heat and particles, if the piston contains holes that the molecules can fit through; and so on. If the exchange isn't too quick, more stuff flows in one direction (say, from the handful to the rest) than oppositely, for a while. Eventually, the handful reaches equilibrium with the rest: stuff flows back and forth, but as much stuff flows in one direction as in the other, on average.

The things exchanged are measurable properties—what we called *observables* in chapter 8. We can measure the handful's energy, the number of particles in the handful, the handful's electric charge, and so on. We might expect numbers to represent quantum observables as they represent classical observables. As a classical human, I can purchase two steampunk novels at a bookstore for $7.84. Thereafter, I may reemerge into the negative

12.4-degree-Fahrenheit outdoors and walk zero steps before dropping a mitten. Two, 7.84, -12.4, and zero are *real numbers*. Real numbers represent the observables in our everyday lives—and represent everyday observables completely. Even when we're not measuring the temperature, a real number represents it.

Real numbers help us represent quantum observables in a more limited fashion: If we measure a quantum observable, the outcome is a real number. But if we're not measuring the observable, it can't be represented with one real number. Attributing just one real number to an observable attributes a well-defined value to that observable. Quantum observables don't always have well-defined values. For instance, the more well-defined a quantum particle's position, the less well-defined the particle's momentum, according to the uncertainty principle. Single real numbers can't capture quantum uncertainty, or the randomness of quantum measurements' outcomes.

Instead, we represent a quantum observable with a bunch of numbers arranged in a square grid. Scientists call the grid a *matrix*, but I'll call it an *übernumber*. Why? First, if your garden contains a Shakespearean garden, a Chinese garden, and a Japanese garden, then your garden is an übergarden, being a garden that consists of gardens. Similarly, an übernumber consists of numbers and acts like a number, as you can add and subtract übernumbers. Second, I want the excuse to use the prefix *über*.*

Übernumbers don't obey all the rules followed by real numbers. For instance, imagine multiplying two real numbers together. The order in which you multiply the real numbers doesn't matter: two times three equals three times two. Real numbers, we say, *commute*

* Such a matrix can contain not just real numbers, but two-dimensional numbers, or complex numbers. How can a number be two-dimensional? Being on Earth, I'm on a two-dimensional surface. We can specify my location by specifying a latitude and a longitude. Those two real numbers can be packaged into one complex number. In general, a complex number contains two real numbers and can be interpreted as a point on a two-dimensional surface.

with each other. Now, imagine multiplying two übernumbers together. The order often matters; not all übernumbers commute with each other. Consequently, quantum physics is littered with jokes of the form: "Why do Professor Jones and Professor Smith live only a block from campus? Because they won't commute!"

Übernumbers' refusal to commute underlies quantum uncertainty mathematically. For instance, übernumbers that don't commute with each other represent position and momentum. We say that quantum position and momentum *disagree* with each other, which makes them sound like neighbors squabbling over a property line. Übernumbers' refusal to commute, or disagreeing observables, distinguishes much of quantum physics from classical physics.

But recall our handful of particles exchanging observables with the bath. Suppose that the particles are not hot air molecules, described by classical physics, but cold atoms. Each observable exchanged—energy, electric charge, and so on—is represented by an übernumber. Physicists usually assume that these übernumbers commute with each other. Almost no one realizes that they make this assumption; it slips past most people's consciousness. But the assumption lurks within mathematics that often accompanies this thermodynamic story. So, unawares, we blind ourselves to a source of nonclassical behavior in thermodynamics. We see only the mundane—commuting übernumbers, or agreeing observables, or metaphorical workers' huts.

What happens if the handful of atoms and the bath exchange observables that disagree with each other? For instance, each atom may have a spin that serves as a qubit. The observables exchanged can be components of the spin: We envision a qubit as an arrow pointing in some direction. We can break a direction into three components: upward or downward, leftward or rightward, and forward or backward. Each component corresponds to one observable: an experimentalist can measure whether a qubit points upward or downward, measure whether the qubit points leftward or rightward, or measure whether the qubit points forward or backward. Each measurement is of one observable, one component of the spin. These observables disagree with each other.

So, what happens to thermodynamics if the observables exchanged disagree? I stumbled upon this question with a collaborator early in my pursuit of a PhD.[1,2] Quantum thermodynamicists in London stumbled upon it simultaneously.[3] One school of physicists had dug the question up decades earlier: the school of Edwin Thompson Jaynes.

Jaynes was a twentieth-century American physicist who crossbred statistical mechanics with information theory. His ghost still haunts Washington University in St. Louis, the red-brick, Midwestern institution where he worked. Jaynes had consulted for a startup, I heard from a quantum thermodynamicist now at the university. The startup lacked the funds to pay him and offered stock as compensation. That company was Varian, which grew into a behemoth of a medical equipment company. So, Jaynes bought a mansion, where he hosted physics professors and students on Fridays. He'd play a grand piano, and guests would accompany him on their instruments. Jaynes's spirit lives on in the department's weekly potluck dinners, as well as in quantum steampunk.

Jaynes united information theory and statistical mechanics in 1957; his two papers on the subject did for quantum steampunk what *Twenty Thousand Leagues under the Sea* did for science fiction.[4,5] The unification centers on the quantum state reached by the handful of atoms upon equilibrating with the bath. That quantum state is, to a thermodynamicist, like the pressure of water to Captain Nemo: knowing the pressure of water, Nemo can predict how his submarine will fare throughout the ocean's depths. Knowing the equilibrium state, thermodynamicists can make predictions testable in experiments. By "know the equilibrium state," I mean, write down mathematics that represents the quantum state, analogously to writing down the number that's the pressure of water. Different thermodynamicists deduce the equilibrium state's mathematical form in different ways. Jaynes couched his deduction in terms of information and ignorance. Naturally, he invoked liver—I mean, entropy.

Jaynes's entropic strategy works, he noted, even if the observables involved disagree with each other. And he didn't note much more. He wrote only one paragraph about this peculiarly quantum kingdom of quantum thermodynamics—of disagreeing observables that a system exchanges with a bath. Jaynes didn't speculate about which disagreeing observables could be exchanged, which sorts of system would exchange such observables with a bath, or which mechanisms could transport such observables. The unseen kingdom remained largely unseen.

A few of Jaynes's acolytes visited the kingdom.[6,7] Some aimed to complement Jaynes's argument by arriving at the equilibrium state via a different track. Their aim was true, but they hit a mathematical roadblock. So, the kingdom returned to obscurity, unseen by the rest of the world for decades.

A KNOCK AT THE GATE

Three cohorts of quantum thermodynamicists (one of the cohorts included me) arrived at the kingdom's doorstep a few years ago.[8–10] We'd traveled different paths; mine began at the federation of resource theories. A collaborator and I had generalized resource theories to model exchanges of all sorts of observables in thermodynamics—heat, particles, electric charge, magnetization—almost.[1,2] Our mathematics held water if the observables agreed with each other. Our results included yet another path to the equilibrium state. So, if the observables disagreed, we couldn't conclude that the small system—the handful of atoms—would equilibrate with the bath.

Jaynes established that the handful would equilibrate, didn't he? Not quite; he described the quantum state with which the equilibration supposedly ended but said nothing about the equilibration process. Furthermore, Jaynes's reasoning relies on information theory and not much physics. So, Jaynes's quantum state might exist only in his imagination, not as the result of any physical process.

We quantum thermodynamicists, in questing for the unseen kingdom, determined to ground this equilibrium state in physics by deducing it in other ways. We set out along a sand-strewn track through the thermodynamics of observables that agree with each other. For instance, suppose that a handful of atoms and a bath exchange atoms of two elements—audron and baxterium—as well as energy. The number of audron atoms in the handful, the number of baxterium atoms, and the energy are represented by übernumbers that commute with each other.

The handful of atoms and the bath, together, form one composite system. Suppose we'd like to prepare the composite so that the handful, upon exchanging atoms with the bath, equilibrates. We'll prepare the composite with a well-defined number of audron atoms, a well-defined number of baxterium atoms, and a pretty well-defined amount of energy.* The handful will then exchange energy and atoms with the bath, equilibrating.

The handful winds up in a quantum state whose mathematical form we want to deduce. We can deduce the form from the composite system's quantum state, which encodes the total number of audron atoms, the total number of baxterium atoms, the approximate total energy, and no other information. We call this composite quantum state a *microcanonical state*. Its classical analog is called the *microcanonical ensemble*, which, a college professor of mine

* The energy is only fairly well-defined because quantum theory distinguishes energy from other observables. We won't worry, here, about why.

pointed out, sounds like what Napoleon would have called part of his army.

So the composite system is in a microcanonical state. If we can rub out all traces of the bath from that state, we'll be left with the handful's state. Does some mathematical operation equate to rubbing out all traces of the bath? One does—glory be to the quantum physicist's shield and sword: linear algebra. Thusly we deduce the few atoms' equilibrium state from the whole system's microcanonical state, if the observables exchanged agree with each other.

What if the observables exchanged don't agree? We steer our horses off the track, onto sand more likely to ensnare their hooves. Let's stick as close to the track as possible. Let's try to follow the microcanonical deduction, supposing that the observables exchanged don't agree, and see what happens.

The handful of atoms has the best chance of equilibrating if we prepare the handful-and-bath composite in a microcanonical state—with a pretty well-defined energy and a well-defined value of each other observable exchanged. Rein in—a well-defined value of each other observable? But these observables disagree with each other; they participate in uncertainty relations as position and momentum do. The observables can't necessarily have well-defined values simultaneously.

Pausing beside the track, we confer, our horses sweating and snorting. After a few minutes, we turn away from the track and ride up a sand dune. Our path bends when the track bends and straightens when the track straightens. But, being farther up, we command a broader view of the landscape than we could on the track.

The microcanonical state constitutes the track, a conventional route to the equilibrium state. Turning off the track, we generalize the microcanonical state: instead of giving each observable (apart from energy) a well-defined value, we enable it to have a somewhat well-defined value. This blurriness enables the disagreeing observables to satisfy the uncertainty principle. Rub out the bath from the near-microcanonical state, and you deduce the few atoms' equilibrium state. It turns out to be close to the quantum state predicted by Jaynes. So, a coalition of quantum physics, information theory, and thermodynamics—quantum steampunk—won the day.

The Gate Creaks Open

More sojourners have joined us in the unseen kingdom. Over the past few years, exchanges of disagreeing observables have rippled across quantum-information thermodynamics. Some physicists have developed a resource-theory model; some have proved second laws; some have considered how you'd store observables that disagree as a battery stores energy. A scientist—a traveler tasked with uncovering new lands—lives for such ripples.

Yet I wanted more. Quantum-information thermodynamics is abstract and mathematical. I adore it, and it's taught us much, but it offers only one perspective. Could any real systems in our world exchange disagreeing observables in thermodynamics? How would such systems look? What would they consist of? Might they exist in nature, and if so, where? Could we craft them in a laboratory, and if so, how? The nitty gritty of experiment would round out the abstract mathematics of quantum information theory.

So, I proposed an experiment, with two fellow travelers. We showed how to observe a small system, akin to the handful of atoms, approaching the equilibrium state kissed by quantum uncertainty.[11] Today's experimentalists have the tools needed to perform our experiment on cold atoms and trapped ions.

We showed how to put ions' spins in a near-microcanonical state—such that, if you measure how many spins point upward—or how many point leftward, or how many point forward—the answer is random but sufficiently predictable. Furthermore, we showed, a handful of ions and its bath can exchange all three observables: upward-pointing-ness, leftward-pointing-ness, and forward-pointing-ness. After the ions exchange observables for a while, the handful ends up near its equilibrium state. A bridge opened, for disagreeing exchanged observables, from quantum-information thermodynamics to atomic physics—from the abstract to the real world.

Bridges assume many forms. The bridge from Audrey's world to the unseen kingdom consists of a letter. The bridge between quantum physics and information theory, I envision as gleaming metal—a modern beauty whose finishing touches we continue to add. My two friends and I built a rope bridge, for disagreeing exchanged observables, from thermodynamics to experiments. Those observables, I believe, deserve wrought iron with twirly bits sticking out—not gaudy twirly bits, mind, but tasteful ones. And lamp posts every few yards, with twirly bits of their own.

Construction on the bridge continues. We're working with experimentalists to implement our experiment with trapped ions in their laboratory. Fingers crossed, I'll report on the outcome in this book's second edition. Colleagues and I are also broadening the bridge to reach not only atomic physics, but also condensed matter and particle physics. Scientists in those fields, too, ponder equilibration in quantum systems. They adopt different approaches than quantum thermodynamicists; and they've developed many tools—mathematical, conceptual, and experimental—for understanding quantum equilibration. I aim to generalize those tools to accommodate disagreeing exchanged observables. We can use the generalizations to study how quantum disagreement affects the transport and storage of energy and information.

We already know a little about effects on transport: disagreeing observables produce less entropy, hopping between quantum systems, than agreeing observables do.[12] The less entropy, the merrier, as producing little entropy roughly amounts to aging slowly. Furthermore, I suspect that disagreeing exchanged observables might help us build memories for quantum computers. So, disagreeing exchanged observables appear useful, in addition to fundamentally distinguishing quantum thermodynamics from classical thermodynamics.

Disagreeing observables in thermodynamics have grown into a cottage industry, and I hope to help grow them into a subdiscipline. The subdiscipline will appear, on the quantum-steampunk map, as a kingdom discovered after remaining unnoticed for decades—like the kingdom, in Audrey's novel, unseen for millennia.

CHAPTER 13

All Over the Map

ROUNDING OUT OUR TOUR

Landscapes flashed past as the travelers walked, crawled, motored, rode, and flew, night and day. Single images impressed themselves upon Audrey's memory, like leaves snatched from a whirlwind: In a tunnel cut from the rock below the Earth's surface, a single torch illuminated the human figures carved on the walls. Rows of huts stretched across a barren landscape broken only by withered shrubs scattered here and there. Blood red, lemon yellow, emerald green, and sky blue clamored for attention in dresses, linens, turbans, and woven baskets in a bazaar. A glass palace glittered atop a cliff under a stormy sky, grey waves crashing against the rocks below.

Everywhere the travelers arrived, Ewart sought them out. In the bazaar, they were stalked by a thin, reed-like man smoking a pipe—always more than a stall away but never far enough. A letter, awaiting the trio at an inn they reached late one night, had been slitted open and resealed before their arrival. Shadows clung to the walls, and clouds didn't hang over only the glass palace.

Every time they arrived at a mail depot, the travelers teleported a few qubits to Audrey's parents, as the quantum information could reach the elder Stoqhardts more securely than classical information could. Much as she reveled in exploring, and in the sense of purpose provided by their quest, and in how her stomach plunged whenever their gyropter dipped, Audrey sometimes envied those qubits. On frigid nights, huddled beneath a thin blanket, as Baxter snored in the neighboring cot, she wished she could teleport like one of those qubits to the travelers' final destination. Alas for Audrey, a classical lady could not teleport—and, worse, none of the travelers could predict where their final destination lay.

WE'VE WALKED, crawled, motored, ridden, and flown across much of the quantum-steampunk landscape. Yet city-states and principalities still lie before us. I'll touch on several briefly, as in a whirlwind tour of Europe. One can't grow acquainted with London, Paris, Rome, and Venice in one week; but one can taste the cities, developing a sense of each region's flavor.

{ Cool It }

Computers, we've established, need blank scrap paper. For instance, imagine calculating the distance traversed by Audrey, Baxter, and Caspian. You'd trace their route out on a map, measure each leg of the trip with a ruler, convert the length on the map into a length on the ground (or in the air), and sum the lengths. We'd write the numbers down, then write more while carrying 1s and so on during the summation.*

Blank scrap paper manifests, to a classical computer, as bits set to 0. To a quantum computer, blank paper manifests as qubits represented by upward-pointing arrows. Imagine arranging magnets near an electron as in figure 13.1. The south pole lies above the north pole, so the magnetic field points upward. The electron's spin interacts with the magnetic field, which shapes the spin's energy ladder. The spin occupies its low-energy rung when aligning with the magnetic field (when pointing upward) and occupies its high-energy rung when anti-aligning (when pointing downward). So, dropping the temperature, lowering the qubit's energy, forces the spin upward. Cooling erases the scribbles from quantum scrap paper.

Erasing scribbles costs work, we learned from Landauer. In other words, the information-processing task of erasure happens

* Some people could perform the computations in their heads, but they'd use a neuronal equivalent of scrap paper.

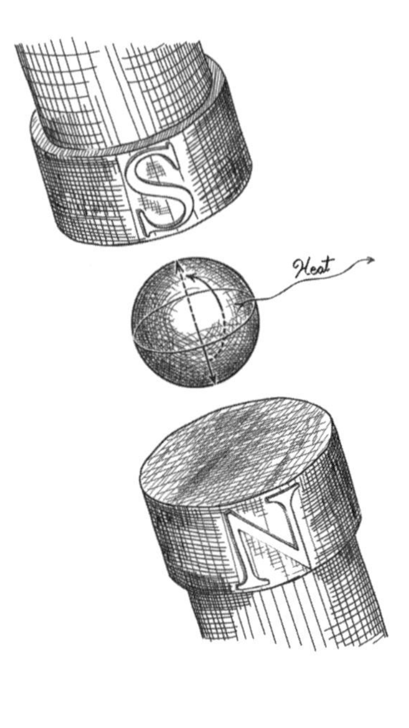

FIGURE 13.1

via a thermodynamic process. Conversely, we can accomplish a thermodynamic process by processing information: we can cool qubits by exploiting correlations between them. We call this strategy *algorithmic cooling*.[1,2]

Imagine a clutch of qubits that we've used to factor 9,824,783 into prime numbers. Individual qubits contain entropy, as a corner of a sheet of scrap paper can contain scrawlings. Some of the qubits might share entanglement; some might share classical correlations. We can design a circuit that shoves the entropy into a few of the qubits. Think of those qubits as a rubbish bin (as Audrey would say) where we dump the rinds and grease from dinner. The rest of the qubits emerge reset to pointing upward, like dishes emerging from a scrub.

This strategy can accommodate one-shot thermodynamics, à

la chapter 10: our next computation will employ the qubits that we'll have cleaned. We might not mind if the next computation has a 0.1% chance of failing, so the cleaned qubits need only point *mostly* upward. We can clean more qubits approximately than exactly; so, by sacrificing the next calculation's probability of succeeding, we can scrub more qubits.

A classical example illustrates algorithmic cooling, which can reset classical bits as well as qubits. Suppose that Audrey holds one bit and Baxter holds another. The bits have the same value, sharing a classical correlation. The pair has a 50% chance of being 00 and a 50% chance of being 11.

One logical operation, akin to addition, is the *controlled-NOT*. If Audrey's bit is a 0, the operation does nothing to Baxter's bit. If Audrey's bit is a 1, the operation flips Baxter's bit, interchanging the 0 and 1 values. We call this flip a NOT operation because 0 often stands for *yes*, while 1 stands for *no*.* *Not yes* means *no*, and *not no* means *yes*.

To understand the controlled-NOT, imagine Audrey and Baxter visiting their great uncle, Lord Wyndham Whewell. The first morning, Lord Whewell's butler asks the siblings whether they'll partake of the soup to be served at dinner, as neither cares much for soup. Baxter, a younger brother who looks up to his sister, decides to answer as she does. So, both their answers—both bits—are 0s (*yes, we'll have the soup*), or both are 1s (*no, thank you*). Audrey waffles, assenting during half their visits and declining during half. So, the bits have a 50% chance of being 00 and a 50% chance of being 11. Ten minutes before dinner, the butler confirms their choices. Audrey doesn't change her mind. If she continues to consent to the soup (if her bit remains a 0), Baxter continues to consent (his bit remains a 0). However, if Audrey continues to decline the soup (if her bit remains a 1), hearing her negative encourages Baxter's caprice momentarily,

* At least, in quantum computation. In other disciplines, 0 means *no*.

and he *changes* his order (he flips his bit from 1 to 0). The controlled-NOT negates Baxter's choice conditionally on Audrey's.

Let's forget the soup and focus on how the controlled-NOT leaves the bits. First, suppose that the bits begin as 00. Since Audrey holds a 0, the operation does nothing to Baxter's bit. The siblings end with 00. Second, suppose, that the bits begin as 11. Since Audrey holds a 1, Baxter's bit flips from 1 to 0. The siblings end with 10.

In both cases, Baxter ends with a 0. His bit has been reset; the next time the siblings run a computation, they can use his bit as scrap paper. But Audrey's bit has a 50% chance of being a 0 and a 50% chance of being a 1. This probability distribution has the greatest entropy possible. So, Audrey effectively holds a "scribbled-on" piece of scrap paper. She discards it, writing her bit off as the price of erasing Baxter's bit.

Baxter's bit ends up clean in this example, but the bit doesn't in all examples. Why? The two bits could begin less correlated, or subject to a probability distribution other than 50-50. The siblings might be able to scrub Baxter's bit only partially, like dishes in the absence of soap.

A heat bath can help.[3] Suppose that Audrey and Baxter have many qubits and a heat bath. The bath is cold but not as cold as their qubits need to be: Thermalizing all the qubits with the bath would erase each qubit partially; each would shed some of its energy and entropy into the bath. But no qubit would shed enough to serve in the next computation; no qubit would arrive close enough to pointing upward.

The siblings evade this pitfall, taking advantage of the bath's low temperature, by alternating steps in an algorithm.[4] First, they perform controlled-NOT-type logic gates on the qubits. This step shifts much of the qubits' energy and entropy into a few rubbish-bin qubits. The rest of the qubits approach the quantum 0 state but don't arrive as close as desired. Second, Audrey and

Baxter thermalize the rubbish-bin qubits with the bath. The bath partially empties the rubbish, accepting energy and entropy from the qubits. These qubits can now accept more energy and entropy from their fellows, during the next logic gates.

The siblings perform logic gates, empty the rubbish bin, perform logic gates, empty the rubbish bin, and so on. When they stop depends on how cold they need their qubits and how cold the qubits can grow. If the siblings play their cards right, the clean qubits can grow colder than the bath. How cold depends on the number of qubits, on the bath's temperature, and on which logic gates the siblings perform.

Relying on logic gates earns this procedure its name, *algorithmic cooling*. Algorithmic cooling has advanced from theory into experiments.[3] Many experiments feature molecules whose nuclei contain the spins that serve as qubits. Experimentalists manipulate these qubits with magnets, applying techniques used to take pictures of the brain in MRI—nuclear magnetic resonance, according to chapter 3.

How quantum is algorithmic cooling? The answer resembles our answer to the question, in chapter 7, "How quantum was the operation of the maser heat engine?" Like the engine, the cooled qubits have quantized energies—energy ladders with separated rungs. But the maser engine needed no entanglement to run. Similarly, entanglement plays no role in most algorithmic cooling. After all, we saw Audrey and Baxter algorithmically cool a bit in a classical example. But algorithmic cooling can prepare qubits to undergo quantum computations that involve entanglement.

{ Uncertainty Relation, v2.0 }

I've never watched a meteor shower. I've never witnessed that first burst of light shooting across the velvety night sky—followed by another, and another, until forgetting the chill seeping through my fleece, forgetting how dry my eyes feel so late at night, and reveling in the light show above. But I can imagine how watching a meteor shower feels. I imagine that it feels like watching a scientific field being discovered, coalesce, and gain momentum. The field of *thermodynamic uncertainty relations* was discovered and coalesced over the past few years. It's now ripping across the sky of statistical mechanics.

We've encountered the quantum uncertainty principle several times. It dictates that the more well-defined a quantum system's position is, the less well-defined the system's momentum is. Quantum uncertainty stems from how quantum observables are represented by übernumbers (matrices) that needn't commute with each other, as we saw in chapter 12. Different origins gave rise to thermodynamic uncertainty relations. So, thermodynamic uncertainty relations govern classical systems, as well as some quantum ones.

But thermodynamic uncertainty relations share parts of their structure with quantum uncertainty relations: A quantum uncertainty relation is an inequality, stating that one number is at least as large as another. On one side of the inequality sit our uncertainties about, for example, a particle's position and momentum. On the inequality's other side sits a number greater than 0. Similarly, a thermodynamic uncertainty is an inequality. We can understand what sits on each side through a story.

We've imagined, many times, two heat baths at different temperatures. For example, given a hot bath and a cold bath, we can extract work via an engine. The engine siphons off energy that flows from the hot bath to the cold. Similarly, one bath may contain particles at a higher concentration than the other bath, or a higher concentration of electric charge. Particles and charge will flow between the baths. Whatever flows—heat, particles, charge, and so on—forms a current.

Thermodynamic uncertainty relations govern currents of all these types. We'll focus on currents of particles, which are easy to

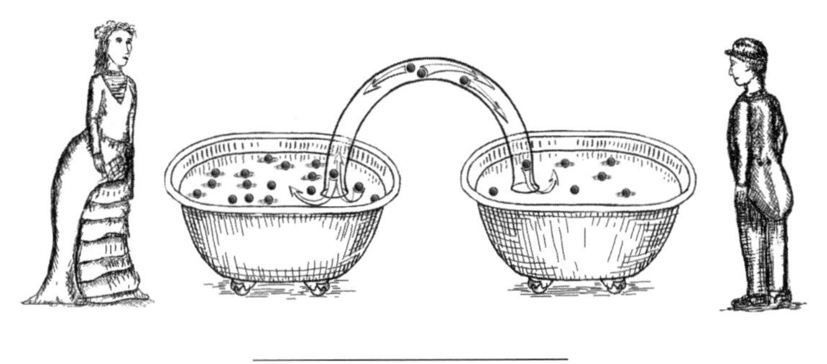

FIGURE 13.2

visualize. Imagine that Audrey holds a high-concentration bath and Baxter holds a low-concentration bath (figure 13.2).

The concentration difference tends to draw particles from Audrey's bath to Baxter's. But life's journey is rarely smooth—for thermodynamic particles as for humans. The particles jiggle around randomly. Some undertake the journey immediately, and some dilly-dally. Some hop from Audrey's bath to Baxter's, then backtrack. So, the current—the number of particles flowing from Audrey's bath to Baxter's—fluctuates from second to second.

Predictability can benefit us. For example, consider the London Eye, the 443-foot-tall Ferris wheel on the River Thames. Electricity propels the wheel. Imagine riding the wheel up, up, up into the sky—topping the circuit, gazing down upon the city spread below your feet—and then feeling a judder. And another judder.

"We apologize for the inconvenience," says a voice over an intercom. "Our electricity is fluctuating."

You might have a smooth ride for the rest of the trip, but you might retrace your steps or you might be flung out over the void by a sudden jerk.

This story illustrates the usefulness of a steady current. It doesn't accurately portray the London Eye. Authorities wouldn't have approved the ride if it posed a serious risk of flinging passengers out over the void. Besides, the London Eye requires lots of current. The average current is so high, the fluctuations are barely noticeable in its shadow. But small amounts of current flow across cell membranes and through cells in your body. At scales that small, fluctuations matter. According to thermodynamic uncertainty relations, the fluctuations can't be too small. The inequalities set a lower limit on our uncertainty about a current, as inequalities set a lower limit on quantum uncertainties.

The lower limit on current fluctuations depends on liver—that is, on entropy. A bath gains entropy whenever a bath gains particles. The entropy produced depends on the number of particles gained, the concentration of particles in the bath, and the bath's temperature. The greater the entropy dissipated, the less the current necessarily fluctuates.

Why? Imagine the current as a cartload of turnips wheeled across a bridge by a young farmer. A crusty, old official monitors the bridge, overseeing commerce. He collects tolls, paid in entropy, for crossing. Occasionally, the farmer brings too few turnips. The official huffs and grumbles about the hardships that will befall his people: families will suffer tasteless soups and thin vegetable pies. In light of the inconvenience, the official demands an extra fee. On other days, the farmer brings too many turnips. Again, the official groans and gripes: What will his people do with so many turnips, use them to stop up chinks in their walls? Again, the official charges a penalty.

The more extreme the departure from the mean turnip delivery, the more entropy the official charges the farmer. So, the farmer cleaves close to the average, most days. As with turnips and fees, so with particles and entropy.

Thermodynamic uncertainty relations are ripping across the sky of statistical mechanics for reasons similar to why fluctuation relations ripped across the sky years ago. Thermodynamic uncertainty relations introduce law and order into the Wild, Wild West far from equilibrium. The relations have withstood tests, and they govern diverse settings.[5]

Example settings include molecular motors, which haul cargo along the scaffolding in biological cells.[6] How efficiently can such a motor operate? The second law of thermodynamics offers little insight, implying only that the efficiency is no better than perfect. A thermodynamic uncertainty relation constrains the efficiency more tightly. This constraint is useful because it facilitates predictions about the motors' operation.

Also, being able to violate such a constraint is useful: if we can achieve an efficiency greater than allowed by a constraint, we can turn any amount of heat into more work than expected. Some quantum systems violate constraints, stemming from thermodynamic uncertainty relations, on engines' efficiencies.[7,8] That some systems violate the constraints shouldn't surprise us: every conclusion rests on assumptions. We deduce thermodynamic uncertainty relations from assumptions about how particles flow between baths. Not all particles flow according to the same rules. But we can accomplish more than predicting that some systems violate the constraints: we can identify which systems do, why they do, and how they do. This identification can help us design engines—perhaps made from quantum dots, or artificial atoms—to meet our needs. Furthermore, physicists tailored new thermodynamic uncertainty relations to quantum systems.[9] The meteor shower has yet to end.

Haste with Less Waste

Quantum thermodynamicists have a reputation for dragging our heels. Not that we hold up traffic on the road or perform calculations slowly, but we like slow protocols. For example, recall Carnot's engine. It operates at the greatest efficiency possible only while running infinitely slowly. As another example, consider

Landauer's principle: Erasing a bit costs at least a szilard of work, we established in chapter 5. We pay the minimum amount possible only if operating infinitely slowly; speeding up dissipates extra work as heat.

Those examples feature only classical costs; quantum physics introduces more. For instance, consider an atom on its energy ladder's second-lowest rung. We may need to keep the atom on that rung while turning on an electric field. One can turn on an electric field by bringing positive and negative charges near the atom. But we can imagine turning on the field by twisting a dial, as we'd turn up the volume on an analog car radio. Twisting the dial changes the ladder: Some rungs may shift upward, and some may shift downward. Some rungs may break apart, swap pieces with other rungs, and be glued back together (figure 13.3). The atom will remain on the second rung if we turn the field on slowly. Such a slow adjustment is called *quantum adiabatic*.

A quick adjustment shoves the atom from its rung onto a superposition of rungs. A superposition is more complicated than one rung is, so we've created a mess, which may ruin our experiment. Also, the superposition can include rungs that have greater energies than the original rung. Giving the atom a foot on a higher rung costs energy. We have to pay this energy as work (or as heat, depending on our definitions of quantum work and heat). In thermodynamics and especially in quantum thermodynamics, haste makes waste.

At least, haste usually makes waste. We can circumvent this maxim using a *shortcut to adiabaticity*. Shortcuts resemble goggles (a necessity in every steampunk costume) in altering our outlook—in training our gaze on two requirements of our adjustment of the energy ladder: First, the atom must begin on its second energy rung, with no electric field present. Second, the atom must end on its second rung, with an electric field present. Aside from those two requirements, anything goes. We've been assuming that we have to

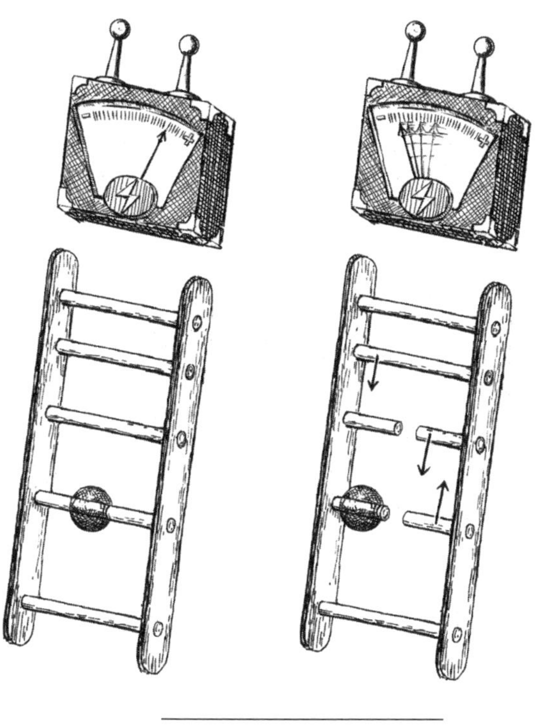

FIGURE 13.3

turn the electric field on steadily, but multiple paths lead to our destination.

For example, we could zigzag—strengthen the field a lot, then weaken it a little, then strengthen it much more, then weaken it a little, till arriving at the desired strength. Or we could add more ingredients. Suppose, for example, that the electric field must end pointing upward: Negative charges must sit above the atom, and positive charges must sit below. We could add an electric field that points sideways, as by adding negative charges on one side and positive charges on the other side. Or we could add magnets, or bring more atoms to interact with the first. We'd have to remove those extra charges, magnets, or atoms, before ending our manipulations. But they demonstrate that our toolbox contains more than one tool.

The extra tools wrench the atom off the second energy rung. They could send the atom high up on its ladder, or far down, or into superposition after superposition. But who cares, if they replace the atom on the second rung and act quickly? Imagine the atom as a boy whose mother dispatches him to visit an auntie living north of his village. The boy mounts his bike and prepares to pedal northward. But an older woman, wearing a singed checkered jacket and goggles pushed up over her flyaway grey hair,* stops him in the street. The woman has invented rocket boosters, which she'll let the boy test-drive. But the boosters, for some reason as kooky as their inventor, push only in directions halfway between cardinal directions. The boy accepts, and the woman ties the boosters to the bicycle. The boy zooms northeastward, then compensates by zooming northwestward, then continues weaving back and forth. The boy avoids the straightest path to his destination. But he arrives, and in less time than the straight path would have taken.

Shortcuts to adiabaticity serve as rocket boosters that refuse to push along cardinal directions. Trips to different aunties require different boosters, which drag us into different ditches and meadows. But we can often design boosters, and a path, that will bring us to any given auntie.

As boosters consume energy, so can shortcuts cost work. How much work depends on which definition of quantum work we choose from our menagerie. We're exploring shortcuts partially out of a desire to extract work via an engine quickly. So, if a shortcut costs work, then we pay work to extract work. Why bother with the shortcut, then? First, the relative values of work and time are subjective. We might be content to pay much work to extract slightly more work, provided that we gain the balance

* The British setting demands *grey* rather than *gray*.

quickly. Whether we extract more work than we pay remains under debate, partially because defining quantum work is fiddly.[10] But, as research into shortcuts continues, they might turn out to cost less than they provide.

Second, thermal machines other than heat engines exist—for example, refrigerators and heat pumps. We use such machines not to extract work but to force heat into a lukewarm apartment. Pumping heat requires work; so if we're using a heat pump, we're already willing to pay energy. We might not mind investing more work to speed up the heating.

Shortcuts to adiabaticity benefit not only quantum thermal machines but also other technologies. For instance, some quantum computers require qubits to stay on their lowest energy rungs while their energy ladders change.[11] Shortcuts can spare us from turning the relevant knobs too slowly. Not only those quantum computers, but all types of quantum computers benefit from qubits unscathed by decoherence. The longer a computation lasts, the more time qubits have to entangle with their surroundings. Shortcuts can speed up computations, preserving qubits' isolation. Also, shortcuts benefit quantum communication and metrology. So, quantum thermodynamicists study shortcuts within a broader community of scientists who do. When one woman invents rocket boosters, the whole village benefits.

{ TORNADO TOUR }

The first time I visited England, my mother and I booked a day-long bus tour. We departed from London in the morning; visited Windsor Castle, Stonehenge, and the University of Oxford; and returned to London after dark. I grew to love Oxford only years later, while lingering for days or months. But we didn't have months to spend on that first visit. So, my mother and I made do with half of an afternoon. So shall we make do, dear reader, with a few pages for the final city-states on the quantum-steampunk map. Our whirlwind tour shall end as a tornado.

What distinguishes quantum physics from classical? Not superpositions, which classical waves can be in. Not discreteness, which classical systems can approximate. If you guessed *liver*—that is, *entropy*—you're on the right track: a quantum system has von Neumann entropy when entangled with another system, and entanglement is nonclassical. But what about entanglement enables it to speed up computations? A property called *contextuality*, at least in certain cases.

Every experiment occurs in some context: certain posters adorn the laboratory's walls; a dropped pencil lies on the ground beside the door; the experimentalist wears a gray woolen sweater knitted by her grandmother. Much of this context seems irrelevant to the experiment's outcome. But some elements that we'd expect to be irrelevant—judging by our everyday, classical experiences—are relevant, in quantum physics. Not dropped pencils and woolen sweaters, but quantum analogs of them. So, we call quantum physics *contextual* and classical physics *noncontextual*.

It's difficult to prove that some physical phenomenon is *nonclassical*—that no classical system could reproduce the phenomenon. But your work is done if you can prove that the phenomenon is *contextual*. Which quantum-thermodynamic behaviors are contextual—provably nonclassical—and which aren't? Physicists are answering this question by finding contextuality—or a lack thereof—in protocols for measuring and extracting work.[12]

Time to get back on the tour bus—or, since we're surveying quantum steampunk, the steam-powered motorcar. We're leaving Windsor Castle for Stonehenge—which consists, in this book, of realistic heat baths. Across conventional thermodynamics, we assume that baths have certain properties: First, they're infinitely large. This assumption permeates even resource theories, which contribute to one-shot thermodynamics, the champion of small systems: we assume that one's environment contains baths of all sizes.

Second, we assume that heat baths have short memories. Consider a small system—say, a few qubits—exchanging heat with a bath. As energy passes between bath and qubits, information about the qubits' quantum state enters the bath, like a swimmer swept out to sea. The sea is so vast, the water has almost no chance of returning the swimmer to their shore. Likewise, the bath is so vast, the information has almost no chance of returning to the qubits. The information gets lost in the bath, so we say that the bath lacks a memory. Third, the bath interacts with the qubits weakly; exchanging a great deal of energy takes a long time.

Not all baths satisfy these assumptions. What if a heat bath is small, returns swimmers to their shores, or exchanges energy quickly? The bath can serve as a resource,[13] the qubits may avoid thermalizing, erasure can cost less work than Landauer predicted,[14] and more.

Now, let's motor from Stonehenge to Oxford; that is, let's move on to quantum thermometry. The zeroth law of thermodynamics implies that thermometers exist: If a spoon held by Baxter is in thermal equilibrium with an almond pudding being eaten by Audrey and with a curry being eaten by Caspian, then Audrey's almond pudding is in thermal equilibrium with Caspian's curry. If the friends know the temperature of Caspian's curry, they can infer the temperature of Audrey's almond pudding. Baxter's spoon serves as a thermometer. Thermometers detect and report temperatures, as I learned when plastic rods were pushed deep under my tongue during childhood.

Determining whether a child should stay home from school requires neither high precision nor small thermometers—nor quantum theory nor the ability to discriminate between low temperatures. But imagine studying how an embryo's development depends on the temperatures at different points in the embryo.

You'd want to detect tiny temperature differences across tiny distances. Or imagine measuring a quantum system's temperature.

Quantum thermometrists study the effects of quantum phenomena on temperature measurements—effects that lead to challenges and benefits.[15] Challenges include a contrast with the zeroth-law story above. There, Baxter's spoon equilibrates with Audrey's almond pudding and with Caspian's curry. Equilibrating—and even nearly equilibrating—takes a long time. Given too long to sit and twiddle their thumbs, quantum systems decohere. So, quantum thermometers may not have time to equilibrate with quantum systems.[16]

Another challenge stems from how measurements disturb quantum systems. The more information a measurement extracts, the more the measurement disturbs the quantum system. How much can you learn about the temperature without damaging the quantum state much?

The advantages of quantum thermometers include the ability to entangle. Preparing a thermometer in an entangled state can enhance the measurement's precision.[17] Or suppose that your thermometer consists of one qubit.[16] It can detect temperatures better if prepared in a superposition of energies. Although classical waves can be in superpositions, the plastic rods pushed under my tongue in childhood couldn't.

Welcome back to London. We've completed our tornado tour of Windsor, Stonehenge, and Oxford—and so our whirlwind tour of the quantum-steampunk map. But Audrey, Baxter, and Caspian's journey hasn't ended, and neither has ours. Centuries ago, explorers feared that roaming too far beyond known boundaries would carry them off the planet's edge. Let's realize their worst nightmares, by stepping off the map.

CHAPTER 14

STEPPING OFF THE MAP

QUANTUM STEAMPUNK CROSSES BORDERS

A plume of grey smoke rose from the east, as though from the pipe of a giant who's finished his dinner of Englishman.

"Akram's signal," Caspian whispered. Although Audrey barely heard the words, the effort cost him, and he struggled to suppress a grimace. "Ewart will arrive within the hour."

Caspian was lying in the dirt beside the map, his chest wrapped with a bandage torn from Audrey's petticoat. Audrey and Baxter were kneeling beside the map, its corners weighed down by three stones and a dusty leather water skin. For several minutes, Baxter had been staring at the map silently and fiddling with the Collection. At Caspian's words, Baxter wrenched the penknife out from amongst the tools and stabbed it into the dirt.

"It *can*-not be here," he said, stabbing again. "We have looked *ev*-erywhere and found *noth*-ing, so it simply cannot ex-*ist*. We have been chasing a *ghost*." A stab accompanied each emphasized syllable.

Audrey had been gazing at the narwhal cavorting on an edge of the map, but she glanced up as Baxter flung the Collection to the ground. She was about to admonish him, when she noticed a valley beyond the stabbed dirt. The valley bordered a collection of twisted, tough-looking scrubs, which she saw over the pile of twigs scattered between Baxter and Caspian's feet. Caspian's head lay on a rolled-up cloak beside a mound of sand, behind which, in the distance, rose a hill like a sated giant's belly. Audrey's gaze drifted from the hill back to Caspian. His eyes, despite being bright with pain, were trained on her.

"Go on," he whispered. Caspian could always tell when she was about to do something.

Audrey held his gaze for a moment—fighting the urge to leap to her feet and flee, fighting the knowledge that she could be home and safe, reading comfortably beside the window in her parents' library—and then she reached over the Peninsula of Pettingroft on the map and picked up the Collection.

"Move back," she told Baxter, before drawing in the ground. She drew around the parchment, incorporating the stabbings in the dirt, the pile of twigs, and the mound of sand.

"Indeed," Audrey said quietly, closing the penknife. "It cannot be here—and neither shall we be here by the time Ewart arrives. This parchment is not all that exists." Putting one hand on the map, she looked up at Baxter. "Hence, we shall leave it and invent the rest. We shall step off the map."

WE'VE TRAVERSED the map of quantum thermodynamics—from town to coast to desert to island—from quantum engines to fluctuation relations to resource theories to thermometry. Where can we step next? I believe that the next frontier for quantum thermodynamics lies outside of quantum thermodynamics. Our discipline borders other disciplines whose practitioners care about quantum physics, information, and energy. These other disciplines include chemistry; condensed matter; atomic, molecular, and optical physics; particle physics; quantum gravity; biophysics; and cosmology. We can cross the borders and meet the neighbors. They'll have problems that we might be able to solve, and they'll have tools that we might be able to leverage. The quantum-steampunk map doesn't end where the parchment depicting quantum thermodynamics ends.

Quantum information scientists reached a similar conclusion years ago. As I prepared to start pursuing my PhD, a senior physicist asked which topic I intended to pursue. I answered, "quantum information theory," and he said, "Isn't that field dead?"

Scientists had solved many of the problems in quantum information theory during the preceding 20 years. Where could I plant myself in an already-sown field? But the discipline was turning over a new leaf: Scientists were exporting quantum information's mathematical, conceptual, and experimental tools. Left with fewer theorems to prove about entropies, entanglement, and resources, they were studying entropy, entanglement, and resources in nature and artificial materials. The quantum-information toolkit unlocked discoveries in condensed matter, computer science, mathematics, chemistry, particle physics, and quantum gravity—not to mention thermodynamics. Quantum thermodynamics can now spread its pollen as quantum information theory did.

Interdisciplinarity is a buzzword: many institutions aim to foster it; some throw money at it; and some claim to champion it. The practice of it is elusive, in my experience of visiting roughly 50 research institutions. When asked how much researchers collaborate across disciplines, people often respond that they wished more collaborations happened. In a small handful of institutions, I needn't ask; I see chemists in the physics department, find papers cowritten by physicists and engineers, and hear about collaborations with mathematicians. Interdisciplinarity is a unicorn, shying away from most civilization.

I can understand why, as an interdisciplinarian who spends little time in the quantum-computation landscape where I grew up. I'm often carrying my quantum-information-thermodynamics toolkit slung over one shoulder, hiking across atomic physics, wading in the sea of chemistry, or weaving through traffic in the sprawling city of condensed matter. I usually know less than everyone else in the room about whatever topic is under discussion. To the others in the room, an interdisciplinarian may sound as though she were born yesterday. But if she doesn't mind—oh, what we can discover together.

We've seen three examples of such discoveries: First, crossing condensed matter with quantum thermal machines gave birth to the MBL-mobile. This quantum engine benefits from the discrepancy between two phases of quantum matter, one thermalizing and one not.

Second, linking quantum thermodynamics to chemistry yielded insights about molecular switches prevalent in nature and technologies. Shining light on a photoisomer gives the molecule the opportunity to switch configurations. With what probability does the molecular switch switch? Modeling the photoisomer in a resource theory—a mathematical model developed in quantum thermodynamics—led to a general, thermodynamic-style bound on the switching probability.

Third, small systems and big baths can exchange heat, particles, electric charge, and things represented by übernumbers that don't commute with each other. Quantum thermodynamics has yielded abstract, information-theoretic insights about the noncommuting stuff. Bridging noncommuting stuff to atomic physics revealed how we could manipulate such stuff in a lab. My research group is extending the bridge to condensed matter and particle physics.

I'm not the only quantum thermodynamicist who crosses borders. I'm grateful to work alongside colleagues who reach into condensed matter, or atomic and optical physics, or quantum gravity. Let's reach across one more border, into black holes.

The Only Name That Contained an Exclamation Point

At the heart of our galaxy lies a black hole, one of the densest objects in the universe. As the US National Aeronautics and Space Administration (NASA) explains, "Think of a star ten times more massive than the Sun squeezed into a sphere approximately the diameter of New York City."[1] Although New York City looks large from a human's perspective, it's teensy from the perspective of a mass greater than the sun. So, black holes can exhibit quantum behaviors.

A black hole tugs nearby matter with a gravitational pull like no other: if you enter it, you have no hope of escaping—even light can't escape. An object that emits no light looks black, accounting for the name *black hole.*

But black holes radiate light in a sense, concluded Stephen Hawking, a University of Cambridge physicist, in 1975.[2] According to particle physics, photons can pop into and out of existence, in pairs. Imagine two photons popping into existence near a black hole. One particle may escape, while the other falls in. Physics behaves oddly near black holes, so the escaping particle carries off a little more energy than we'd expect. As compensation—to keep the total amount of energy constant—the second particle carries a little negative energy into the black hole. So, the black hole's energy decreases.

The escaped particle is in a quantum state we're familiar with: The particle looks as though it's in thermal equilibrium with some heat bath. The thermal state has a high entropy if the black hole has a high temperature. If we measure the particle in any way, the outcome will be almost totally random. The particle carries little information.

Suppose that Audrey throws a diary into the black hole. She wrote a secret in the diary, and she doesn't want for Baxter to access the secret. The secret, she reasons, will stay safe in a black hole, from which nothing can escape.

But photons sort of escape, we've seen. Baxter collects the photons as the black hole shrinks. (He'd have to wait for many times his lifetime, but never mind that inconvenience.) The black hole may shrink all the way to nothing. What happened to the information in Audrey's diary? Many physicists believe that information can't disappear; it can only disperse, like a river subdivided into more and more streams, then rivulets, then trickles. You might protest that information disappears whenever we erase pencil marks. But the penciled information remains in the eraser shreds and in the positions of air molecules displaced by the erasing hand. What happened to Audrey's secret? It isn't in the black hole, which no longer exists. Baxter's photons, being thermal, contain little information.

Such is the black-hole information paradox. Progress on it has advanced and plateaued, advanced and plateaued, for decades. Over the past few years, quantum information theory has fueled several advances. So, what happens when Audrey casts her diary into the black hole?

Imagine the black hole as a New York City–sized clump of quantum particles struck by the diary, which manifests as a qubit (if Audrey's secret was very small). The diary injects a little quantum information into the clump. The information spreads across the clump rapidly. Audrey's secret ends up distributed across all the particles, in entanglement that they share. Imagine trying to recover the secret by measuring the particles with a detector. Realistic detectors tend to be able to access only a few particles at once. No measurement of any handful of particles can recover Audrey's secret. The secret doesn't reside in *this* handful, or in *that* handful, or in *that* handful—or in the sum of those handfuls. The secret resides in the relations among all the handfuls, in entanglement across many particles. We say that the information has—or that the particles have, or that the particles' state has—*scrambled*.

Scrambling has roiled the fields of black-hole physics and quantum computation. Alexei Kitaev, the quantum-computing physicist whom I worked for as a teaching assistant, kicked off the craze in 2015.[3] He identified a number that indicates whether particles are scrambled.* If the number is low, we expect information not to have spread out in many-particle entanglement; if the number is high,

* Alexei found this number in a condensed-matter paper published in the Soviet Union during the 1960s.[4] As we said in chapter 3, whatever you discover in certain subfields of physics, a Soviet journal probably published a paper about it between the 1960s and 1980s.

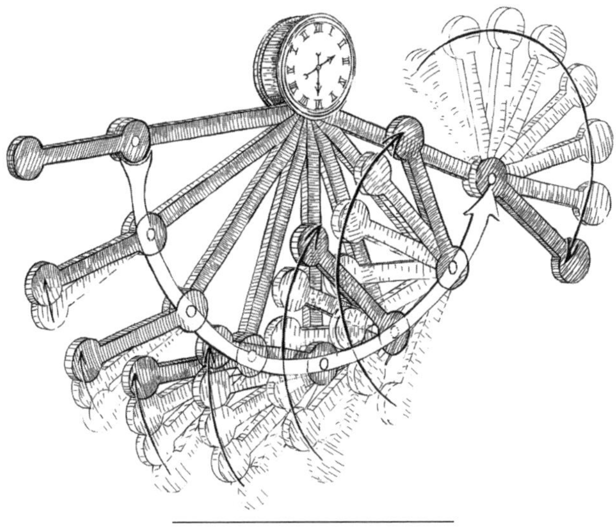

FIGURE 14.1

we expect that information has. The number has a fusty-sounding nine-syllable name—the out-of-time-ordered correlator—so I'll replace the name with *the scrambling signal*. The scrambling signal interrelates scrambling with chaos. Chaos isn't just what reigns the night before a family sets off on vacation ("This suitcase has a hole in it!" "Where's my bathing suit?"). Chaos is a field of physics and mathematics. When I was an undergraduate, my college's mathematics department offered a course about chaos—the only course whose name contained an exclamation point: "Chaos!"

Suppose that Audrey has a double pendulum—a pendulum that hangs from the bottom of another pendulum that hangs from, say, a clock face (figure 14.1). She pulls the bottom pendulum far to her right and then releases it. The double pendulum swings, bends, and loops-the-loop like a trapeze artist. Audrey waits for a while, then photographs the double pendulum (using a high-speed camera available during the Victorian era only in a steampunk novel).

Imagine that Baxter pulls another double pendulum a hair's

breadth farther than Audrey did. He lets the pendulum swing, waits for the same amount of time, and then photographs his double pendulum. Let's compare the two photographs. Baxter's double pendulum is probably in a different configuration than Audrey's. Also, Baxter's was probably moving with a different speed than Audrey's, in a different direction, when the photographs were taken. That is, a double pendulum's motion changes loads if the initial conditions change slightly. This sensitivity to initial conditions characterizes chaos.

Chaos manifests in systems classical and quantum. Classical chaotic systems include the weather. Meteorologist Edward Lorenz encapsulated sensitivity to initial conditions in the term *the butterfly effect*: "a butterfly flapping its wings in Brazil could set off a tornado in Texas."[5] Quantum systems exhibit sensitivity less straightforwardly than the weather, as they obey different equations—quantum theory, rather than classical mechanics. So, detecting quantum systems' sensitivity to initial conditions is tricky. But the scrambling signal provides such a detector.

I count myself fortunate to have had the story from the horse's mouth. Alexei Kitaev taught me about scrambling at the whiteboard in his office, as we planned the final term of our quantum-computing course. He was debating between teaching about black holes, about a mathematical problem, or about another application of quantum computers. I voted for black holes, so he treated the class to an introduction. Then, a piston slid into place in my mind: scrambling belongs with quantum thermodynamics.

Go Scramble Yourself

I developed a sense that I should accomplish something with scrambling, but I couldn't figure out what. So, I dropped in on my PhD advisor's office, and we kicked ideas around. About half an hour in, he dropped a comment that transformed my research and my life: "Well, you're interested in fluctuation relations, right?" Another piston slid into place in my mind.

Fluctuation relations are kin to the second law of thermodynamics, which governs how information spreads out and dissipates. Quantum information spreads out and dissipates through entanglement during scrambling. Furthermore, fluctuation relations compare forward processes to their reverses: Crooks' theorem governs the stretching of a DNA hairpin and the collapse of the hairpin. The scrambling signal, too, encodes forward and backward motions through time.

We can understand how by returning to Baxter's double pendulum. Like Audrey, he's pulled the pendulum far to the right but a hair's breadth more than she has. Baxter releases the pendulum, lets it swing for a while, and stops it after the agreed-upon time. Imagine that he then reverses time, as though pressing *rewind* on a video. (Baxter can't actually reverse time, of course. But he can approximately simulate a reversal of time in an experiment.* Besides, Baxter can write down mathematics that represents a perfect time reversal. So, we'll pretend that he can reverse time in his experiment.) The pendulum ends where Baxter started it (left-hand side of figure 14.2).

Now, let's return to Audrey's double pendulum. She's pulled it far to the right, let it swing for a while, and stopped it. Audrey now nudges the pendulum a hair's breadth rightward, just as Baxter nudged his pendulum before letting it swing. Now, Audrey reverses time, making the pendulum swing backward. You might expect the

* Imagine Baxter experimenting on an electron spin, rather than a pendulum, for simplicity. He can control the spin's evolution—how the spin's quantum state changes in time—with a magnetic field. Suppose that a magnet's north pole lies on one side of the spin and another magnet's south pole lies on the opposite side of the spin. Baxter effectively reverses time's arrow for the spin by reversing the magnetic field—by swapping the north and south poles. No experimentalist can reverse time's arrow for the atom perfectly; the south pole will end up slightly to one side, or the spin will have decohered a tad, and so on. But many a cosmetician would envy Baxter his ability to reverse time approximately.

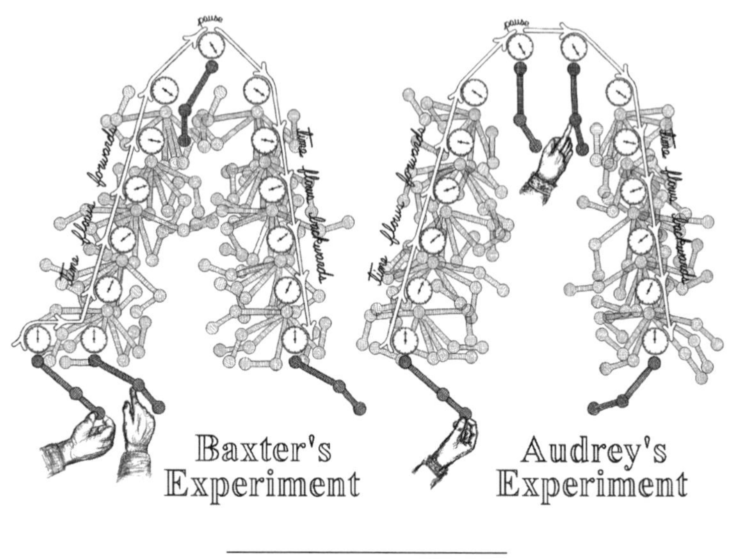

FIGURE 14.2

pendulum to end where Baxter's pendulum ended: Baxter nudged his pendulum, and Audrey nudged hers—just, at different times. But the chaotic motion of Audrey's pendulum amplifies her nudge. Audrey's pendulum likely ends wildly differently from Baxter's: in a different position, with a different momentum (right-hand side of figure 14.2).

A nudge, a forward evolution, and a backward evolution ends wildly differently from a forward evolution, a nudge, and a backward evolution, in the presence of chaos. Be careful what you wish for, storytellers warn us. Be careful when you nudge, chaos admonishes us. Chaos amplifies tiny discrepancies into diverging life paths. We can detect such divergences—and so chaos—by studying time reversals.

Audrey's and Baxter's double pendulums are classical. But we can tell a similar story about quantum systems. The scrambling signal registers the analog of how far Audrey's pendulum ends from Baxter's. So, scrambling is intertwined with time reversal, as fluctuation relations are. Scrambling and fluctuation relations smell the same; they must, I thought, be related somehow.

I found out how.

I determined to prove a fluctuation relation that contained the scrambling signal. The following weekend saw me bound to my desk—writing, erasing, rewriting, obsessed. Within four days, I'd proved something that interested me. With feedback and revision, I proved something that interested my advisor. The fluctuation relation for scrambling[6] kick-started the integration of scrambling and quantum thermodynamics, which have grown closer since. Also, the fluctuation relation kicked off a research program that's led me into black holes, superconducting qubits, the theory of measurements, the theory of probabilities, and metrology—all over and off the map.

The fluctuation relation for scrambling shares its shape with Jarzynski's equality. On one side of Jarzynski's equality sits something useful, something we want to measure: a Boltzmann balance. On the other side sits something we can measure: the probability that, in the next experimental trial, unzipping the DNA will cost an amount W of work. Similarly, on one side of the scrambling equality sits something useful, something we want to measure: the scrambling signal. On the other side sits something we can measure: a quantum variation on a probability. I'll explain the variation momentarily. For now, suffice to say that you can measure the probability-like thing, by exercising some creativity. The scrambling equality's right-hand side even shares its mathematical structure with the right-hand side of Jarzynski's equality.

Jarzynski's equality is useful partially because it offers a means of measuring a Boltzmann balance that's difficult to measure otherwise. The scrambling signal, too, is difficult to measure: It isn't an observable that we can measure directly. It isn't a probability inferable from data about many trials. How to measure the scrambling signal was far from obvious; only about three schemes had been proposed. I discovered an alternative, with help from a friend.

Recall that weak measurements formed the hummingbird definition of quantum work, in our menagerie of definitions. My scheme for measuring scrambling weakly turned out to resemble a scheme for measuring work weakly.[7] The scrambling equality doesn't involve work. But work is a random number that you can measure in an experiment, a number that fluctuates across trials. The scrambling equality similarly involves a random number that you can measure in the experiment I proposed, a number that fluctuates across trials.

This number could replace work in quantum thermodynamics: At the end of chapter 7, an atomic experimentalist refused to measure the work performed by a quantum engine. He could draw all the power he needed from a wall socket, so why care about a quantum engine's minuscule offering? Perhaps thermodynamics should broaden its view beyond work as it broadens beyond classical physics to quantum. Even the experimentalist cares about scrambling, as explained below. The scrambling equality suggests a new cousin of work that's suited to the intersection of thermodynamics and quantum computing.

The scrambling equality differs from Jarzynski's equality not only in replacing work but also in replacing the probabilities with quantum variants of probabilities. The latter replacement makes sense: Jarzynski's equality governs systems quantum and classical. Classical systems don't scramble: Scrambling involves entanglement, as well as uncertainty. So, promoting the probability to a quantum variant, en route from Jarzynski's equality to the scrambling equality, makes sense.

Scrambling has spilled from black-hole physics and quantum computation into other disciplines. For instance, quantum thermodynamicists have now studied entropy production in scrambling.[8,9] Atomic physicists have studied how to mimic a black hole's scrambling in the lab. Condensed-matter physicists have sought the simplest mathematical model of a black hole's interior. Quantum information theorists have linked the scrambling signal to (of course) entropies. Scrambled states underlie a task that today's quantum computers have performed much more quickly than classical supercomputers.[10]

Like textile manufacturing during the Industrial Revolution, scrambling has grown into a cottage industry. It's a subfield whose emergence I've not only watched but also contributed to. I remain grateful for the chance to do both, especially as the fluctuation relation for scrambling has led me farther afield—into metrology, superconducting-qubit experiments, photonic experiments, and interpretations of quantum theory. That's why I see quantum steampunk's future far from the origins of quantum thermodynamics. Stepping off the map requires faith, but it can pay dividends.

EPILOGUE

Where to Next

THE FUTURE OF QUANTUM STEAMPUNK

"Go on."

"Is it safe?" Audrey paused with one hand on the brass tube and one hand suspended above it. Baxter flapped both his hands at his sister, like a duck shaking its wings at a duckling who inquires about a pond's temperature before waddling in.

"Of course, the mechanism is safe," he said. "I checked it half a dozen times, and I promise that it shall not blow our heads off—" Audrey began lowering the suspended hand—"with high probability."

"Baxter!"

"Excuse me." A low voice accompanied the appearance of the Stoqhardts' butler in the doorway. As neither sibling heard him, Caspian beckoned him over to the couch from which Audrey had forbidden Caspian from rising. The doctor had concluded that Caspian would heal—probably, mostly—and Audrey had sworn to fuss over him until he finished.

"Audrey—Ree-ree." Lowering his voice, Baxter used the name he'd called his sister as a toddler. "Of course, the mechanism is safe. I would never let anything happen to you, I promise—with a probability very close to one indeed."

Audrey rolled her eyes.

"The president of the Royal Society has arrived," Caspian interrupted. The siblings turned toward him, Audrey's right hand still suspended in the air over the brass tube. "The rumored invitation to present a lecture about our findings appears likely to become reality. No doubt, however, she expects an informal lecture this morning."

Baxter looked at Audrey, Audrey looked at Caspian, and Caspian raised his eyebrows at Audrey. She looked down at the brass tube in front of her, then turned to the butler.

"Arnold, show Madame Bancroft into the drawing room. Tell her that we shall join her momentarily, and have Daisy prepare almond cake."

"Very good, miss." Arnold bowed and left the library.

"Kind of the Royal Society to express an interest," Audrey said. "Alas—" she looked down at the tube—"the discoveries of today have little patience for those of yesterday." The tube belonged to what looked like a kaleidoscope, and her suspended hand held what appeared to be a plate of multicolored glass. Audrey slid the plate into the kaleidoscope, bent down, and placed an eye to the eyepiece.

LIKE AUDREY, Baxter, and Caspian, we've traversed our map and left it behind. We've visited quantum physics, information theory, and thermodynamics to get our bearings. The three fields coalesced, drawn together by probabilities, entropies, and the abilities of information and work to serve as resources. Distinguishing between classical and quantum thermodynamics, we visited the menagerie of definitions of quantum work and heat. We then steamed along on a quantum engine, battled seasickness in the bay of fluctuation relations, rose through the atmosphere to the confederacy of resource theories, and unearthed a kingdom unseen for decades. Leaving the security of charted land, we struck out from quantum thermodynamics into other fields of science. Entropy has accompanied us throughout the trip, like a Cavalier King Charles Spaniel trotting by our side. I hope that you never look at liver the same way again.

Where will quantum steampunk journey next? I see four frontiers that demand exploration. First, the city-states, towns, and principalities have begun to unify. For example, fluctuation relations have connected to one-shot thermodynamics. We can build more bridges, highways, and byways. They'll unify the subdisciplines within the field, enabling one community to import tools and solutions from another.

Second, we can build more bridges to the world beyond

quantum thermodynamics. Chemistry; condensed matter; atomic, molecular, and optical physics; biophysics; particle physics; and black holes involve quantum physics, information theory, and energy. We've begun to interface with them, as in the fluctuation relation for scrambling, and the implications have redounded across quantum physics. We can reach out, find parallels between fields, and trade more ideas, insights, and questions.

Third, much of quantum thermodynamics consists of theory. A tide of experiments has grown over the past several years, and I expect it to surge. We have quantum computing, among other influences, to thank: the hunger for quantum computers has driven experimentalists to enhance their control over quantum and other small systems. Atoms, ions, photons, superconducting qubits, and nuclei offer platforms for testing quantum thermodynamics.

The opportunity should spur us theorists to propose experiments whose outcomes we can't predict—questions that require experiments. Some quantum-thermodynamics experiments confirm predictions that don't need checking. These predictions rest on quantum theory that has survived tests for decades. Some experiments on which I've collaborated fall into this category. Such experiments advance science by forcing experimentalists to hone techniques, by stimulating theoretical developments, and more. But experiments that we can't easily simulate on classical computers would do more justice to the platforms at our disposal, as well as to our thermodynamic and quantum forbears, some of whom leaned on experiments to uncover theory.

Fourth, we can do justice to our thermodynamic forbears by inventing technologies worth investing in. Like the early thermodynamicists, we've uncovered fundamental physics: we've strengthened the second law of thermodynamics, identified thermodynamic tasks performable with quantum resources but not with classical, and more. But thermodynamics evolved hand in hand with the steam engine, the driver of the Industrial Revolution.

Quantum thermodynamics has the opportunity to drive changes of its own.

Quantum thermodynamicists have proposed technologies: quantum engines, refrigerators, ratchets, and batteries. Such a quantum technology may achieve an efficiency, or a power, or some other metric, forbidden to classical technologies. These discoveries have illuminated the distinction between classical and quantum physics. But are the technologies practical? Not yet. This observation applies foremost to research of mine: for now, I'd rather ride in a classically powered car than in a car powered by the MBL-mobile I coinvented.

Theory contacts technology through experimentation, and experiments have begun. So far, they're proof-of-principle experiments, demonstrating that theory can meet reality after toil and sweat. Most quantum engines require more work—to cool down particles, turn magnetic fields on and off during the engine cycle, and so on—than the engines produce. Autonomous quantum thermal machines may help resolve this problem but remain in their infancy. Colleagues and I aim to nurture them into solutions. I hope that we discover niches suited to quantum thermal machines and closed to classical. In analogy, the strong sunlight in Pasadena, California, suits solar panels as it doesn't suit a power plant. We may identify settings, akin to Pasadena sunlight, that call for quantum thermal machines as Southern California calls for solar panels.

Quantum thermal machines are objects. Techniques developed in quantum steampunk aren't objects, but they can benefit technology, too. Such techniques include shortcuts to adiabaticity, algorithmic cooling, and thermometry. For instance, along with quantum thermodynamicists, other scientists develop and use shortcuts to adiabaticity. Shortcuts discovered to aid quantum engine cycles can also benefit quantum computing, metrology, and communication.

Algorithmic cooling stems from the need for clean scrap paper in quantum computing. Theorists have proposed cooling protocols, and experimentalists have implemented some. To my knowledge, the experiments remain proof-of-principle and showcase algorithmic cooling, rather than using algorithmic cooling to aid experimentalists. But algorithmic cooling may become a screwdriver—a tool that will so appeal to experimentalists that they'll apply it in every experiment. Until then, algorithmic cooling illuminates the relationship between information and heat from a fundamental perspective.

Quantum thermometry has boomed over the past several years. Quantum thermodynamicists have established general principles, illustrated with specific models, of quantum thermometry. Other scientists have been applying quantum phenomena to aid thermometry, too, outside of quantum thermodynamics. For example, two experimental groups applied quantum thermometry to a question in biology:[1,2] Different cells in an embryo divide at different times. Imagine reversing the order in which the cells divide. Would the reversal harm the organism? You could find out by manipulating the temperatures in different parts of the embryo because temperature controls the rate at which cells divide.

Experimentalists injected nanoscale diamonds into a worm embryo. The diamonds reported the temperature at various points in the worm. This information guided experimentalists who heated the embryo with lasers. The manipulated embryos grew into fairly normal adults. But their cells, and their descendants' cells, cycled through the stages of life slowly. Reversing the order in which cells divide seemed not to harm the organisms, despite slowing the organisms' aging. Quantum thermometers illuminated a question in biology.

Yet the embryo experimentalists belong to the field of quantum sensing, off the quantum-steampunk map. The embryo experimentalists have little contact with the town of quantum thermometry in quantum thermodynamics. Similarly, atomic physicists have been cooling quantum gases for decades. The physicists have been taking the gases' temperatures since long before quantum thermometry began booming, and quantum thermometry has not yet infiltrated those experiments. However, quantum thermometry looks poised to surpass standard atomic-physics techniques within a few years. Quantum-thermometry models have grown detailed and platform-specific, and they have the potential to change standards.

These frontiers mark opportunities for quantum steampunk. They beckon, to my mind, like the seas and jungles faced by any adventurer in a steampunk novel. Quantum steampunk has flourished over the past decade: We've gained fundamental insights, such as how quantum resources can outperform classical resources in thermodynamic tasks. We've translated theoretical proposals into experiments, and we've partnered with other fields of science. I expect quantum thermodynamics to continue thriving and evolving. Where the past meets the future, as when thermodynamics meets quantum computation, science can spin a today fit for a steampunk novel.

ACKNOWLEDGMENTS

I'm grateful to many people for their contributions to this book. Thanks to my husband for baking the muffins that fueled my eight o'clock writing sessions on weekend mornings, for all the care and consideration symbolized by those muffins, and for understanding about my need to write at eight o'clock on weekend mornings. Thanks to Sarah Siegel for remaining confident, since middle school, that I'd publish a book.

Thanks to my editors—Tiffany Gasbarrini, Michael Zierler, and Susan Matheson—for their patience and their enthusiasm about the manuscript. Thanks to Todd Cahill for transforming my poor sketches into works of art. This project was supported by grant number FQXi-MGB-2009 from the Foundational Questions Institute and Fetzer Franklin Fund, a donor-advised fund of Silicon Valley Community Foundation. The grant was secured with the help and kindness of Jeffrey Bub. Thanks to Caltech's Institute for Quantum Information and Matter for further support.

Many colleagues and friends dedicated time and attention to reviewing parts of the text and providing feedback: Chris Akers, David Arvidsson-Shukur, Gian Paolo Beretta, Felix Binder, Sara Campbell, Chris Jarzynski, David Jennings, Jay Lawrence, David Limmer, Fred McLean, Jonathan Oppenheim, Jukka Pekola, Patrick Potts, John Preskill, Paul Skrzypczyk, Aephraim Steinberg, and Albert Ye. I owe to Rob Spekkens my awareness of how operationalism likens thermodynamics to information theory. I owe to Jason Alicea the illustration, via a flock of birds, of how more is different. Thanks to Raj Katti and Hengyun Zhou for further technical assistance.

Captain Okoli is named after the late physicist Dr. Chiamaka Okoli, who is missed by her friends and colleagues.

GLOSSARY

ABSOLUTE ZERO. The lowest temperature conceivable. Temperature of zero in units of Kelvin.

AEOLIPILE. Ancient Greek steam engine.

ALGORITHMIC COOLING. Lowering bits' or qubits' temperatures by manipulating correlations among the bits or qubits.

ANGULAR MOMENTUM. A quantity, like energy, possessed by every object that's rotating about an axis. How much angular momentum an object has depends on the object's mass, its speed, and how far each chunk of it lies from the axis.

AUTONOMOUS QUANTUM CLOCK. Autonomous quantum thermal machine that keeps time.

AUTONOMOUS QUANTUM THERMAL MACHINE. Thermal machine, described by quantum theory and not by classical physics, that operates independently, without external control.

BIT. Basic unit of information. The information you gain upon learning the outcome of an event that could have played out in two ways with equal probabilities. For example, the information you gain upon learning how a fair coin landed after being flipped.

BLACK HOLE. Cosmological object so dense that not even light can escape its gravitational pull.

BOLTZMANN BALANCE. Difference between two free energies, such as the free energy of a DNA hairpin long after it's been stretched and the free energy that the hairpin had before it was pulled. Useful number applied in biology, chemistry, and pharmacology. Can be estimated with help from Jarzynski's equality or Crooks' theorem.

BOLTZMANN'S CONSTANT. Number hardwired into our universe, similarly to the electron's mass. Crops up throughout thermodynamics.

BOSON. Fundamental quantum particle that carries force, such as the electric force that attracts electrons to protons. Bosons tend to clump together.

BRUTE FORCE. Simple but time-consuming strategy for solving a computational problem: formulate every possible solution and check whether it's correct, one after the other.

BUTTERFLY EFFECT. Encapsulates a chaotic system's sensitivity to initial conditions. Term coined by meteorologist Edward Lorenz: "A butterfly flapping its wings in Brazil could set off a tornado in Texas."

CARNOT EFFICIENCY. Greatest efficiency achievable by any heat engine that interacts with exactly two different-temperature heat baths during its cycle.

CARNOT ENGINE CYCLE. Engine cycle devised by the nineteenth-century French engineer Nicolas Léonard Sadi Carnot.

CHAOS. Extreme sensitivity to initial conditions. Exhibited by the weather, double pendulums, and black holes.

CIS CONFIGURATION. "Closed" configuration of a photoisomer, or molecular switch.

CLASSICAL. Described accurately by classical mechanics, classical electrodynamics, or general relativity. Quantum physics isn't classical.

CLASSICAL MECHANICS. The physics of objects big enough to be seen with the naked eye, or under a classroom microscope, and how they move. Established by Isaac Newton during the 1600s.

CLOSED, ISOLATED SYSTEM. Thermodynamic system that exchanges nothing (no heat or particles or anything else) with the rest of the world.

COLD BATH. Low-temperature bath that interacts with a heat engine during part of an engine cycle.

COMMUTE. Numbers commute (multiplicatively) if the order in which you multiply them together doesn't matter. For example, two times three equals three times two; so two and three commute.

CONDENSED MATTER. Physics of solids and liquids.

CONTEXTUALITY. Property of entanglement that enables it to speed up computations, in at least some cases. Repeating an experiment many times yields many outcomes that obey some type of statistics. If an experiment's statistics depend on the context in which the experiment was performed—on anything that happened in parallel with the experiment—the experiment is contextual. Quantum theory is contextual; classical physics isn't.

CONTROLLED-NOT. Logical operation performable on two bits (or qubits). If the first bit is a 0, nothing is done to the second bit. If the first bit is a 1, the second bit is flipped (the second bit is changed to a 1 if it began as a 0, or it is changed to a 0 if it began as a 1).

CORRELATION. Property of two (or more) measurements that are performed in many trials. Two measurements share a correlation if, when one measurement's outcome changes, the other measurement's outcome changes.

CROOKS' THEOREM. Fluctuation relation that enables us to predict the likelihood that a given trial will cost (or yield) a given amount of work.

DATA COMPRESSION. The information-processing task of squeezing a message into the least possible number of bits.

DECOHERENCE. Undesirable, uncontrolled entanglement of a quantum system with its environment.

DEFECT IN DIAMOND. Arrangement of atoms that colors a diamond black and enables the diamond to store quantum information. Advisable in quantum computers, not in engagement rings.

DETERMINISTIC. Able to be predicted with certainty. Contrasted with *probabilistic*.

DILUTION REFRIGERATOR. Device used to cool superconducting qubits to near absolute-zero temperature so that they exhibit quantum behaviors.

DISSIPATION. Seeping of energy from a controlled system into many uncontrollable systems. Waste.

DISTURB, DISTURBANCE. See *measurement disturbance*.

DNA HAIRPIN. Length of DNA that consists of two complementary chains connected at one end by a loop. The unzipping and rezipping of a DNA hairpin obeys fluctuation relations.

ELECTRODYNAMICS. Theory of light and its interactions with matter. Developed by James Clerk Maxwell and other scientists during the 1800s.

ENCRYPTION. Security protocol intended to encode information in a form indecipherable to eavesdroppers.

ENERGY. Subject of thermodynamics.

- **CHEMICAL ENERGY.** Energy stored in chemical bonds between atoms.
- **ELECTRICAL ENERGY.** Energy of repulsion between a positive electric charge and a negative charge, as well as the energy of attraction between two like charges.
- **GRAVITATIONAL POTENTIAL ENERGY.** Energy accrued to a mass that resists another mass's gravitational pull.
- **KINETIC ENERGY.** Energy of motion.

ENGINE CYCLE. Sequence of steps undergone by an engine to perform work. Returns the engine to its initial conditions.

ENTANGLEMENT. Relationship shareable by quantum particles. Measurements of the particles can be correlated more strongly than any correlations producible with just classical particles.

ENTROPY. Measure of uncertainty about how an event, such as a measurement, will unfold. Function of surprisals. Many entropies have been defined.

- **ONE-SHOT ENTROPIES.** Functions of probability distributions or quantum states. They measure the best efficiencies with which we can perform information-processing tasks or thermodynamic tasks in a few trials or with few pieces of information.
- **RÉNYI ENTROPIES.** One-shot entropies defined by the twentieth-century Hungarian mathematician Alfréd Rényi.
- **SHANNON ENTROPY.** Measure of the randomness of an event whose possible outcomes are described by a set of probabilities. It equals the best efficiency

with which we can compress classical information, on average, if we compress infinitely many messages.

VON NEUMANN ENTROPY. Measure of the randomness of a quantum state. It equals the best efficiency with which we can compress quantum information, on average, if we compress infinitely many messages.

EQUILIBRIUM. Situation in which a thermodynamic system's large-scale properties (such as temperature and volume) don't change much and no net flows (for example, of particles) enter or leave the system, for a long time.

FERMION. Fundamental quantum particle. Fermions form the matter in our everyday world and obey Pauli's exclusion principle.

FLOQUET DRIVING. Periodically whacking a system. Whether the energy imparted by the whacking counts as heat or as work is debated by quantum thermodynamicists and condensed-matter physicists.

FLUCTUATION RELATION. Equation that's stronger and more detailed than the second law of thermodynamics.

FLUCTUATION RELATION FOR SCRAMBLING. Equation, analogous to Jarzynski's equality, that describes quantum chaos.

FREE ENERGY. The work required to create a system from scratch, as by pulling a rabbit out of a hat, and warming the system up to room temperature. Also, how much work could be extracted by annihilating the system. Features in fluctuation relations.

FREE OPERATION. An operation that can be performed easily, without the payment of any cost, in a resource theory.

FREE SYSTEM. An object that can be accessed easily, without the payment of any cost, in a resource theory.

FUNCTION. Mathematical machine that takes in numbers and spits out numbers. Examples include the logarithm.

GENERAL RELATIVITY. Physical theory that describes large, massive objects, such as the planets. Developed by Albert Einstein during the early 1900s.

HEAT. Random, uncoordinated energy that's transferred between systems.

HEAT BATH, HEAT RESERVOIR. An enormous system that's in equilibrium, that has a fixed temperature, and that may exchange heat with other systems. Soap and back scrubber not included.

HEAT CAPACITY. Amount of heat required to raise a system's temperature by one degree.

HEAT ENGINE. Device that turns heat into a little work while dissipating more heat.

HOT BATH. High-temperature heat bath with which an engine interacts during an engine cycle.

INFORMATION. Ingredient required for one to distinguish between alternatives. Also, that which catalyzes an event without losing its ability to catalyze that event.

INFORMATION THEORY. Study of how to measure information and of how efficiently we can process information (solve computational problems, secure information, communicate information, and store information).

JARZYNSKI'S EQUALITY. Fluctuation relation that interrelates a Boltzmann balance and the work invested to jolt a system out of equilibrium.

LAMASSU. Winged bull-men who guarded the entrances to ancient Assyrian palaces.

LANDAUER ERASURE. The erasing, or resetting, of a bit of information. Costs at least a szilard of work.

LANDAUER'S PRINCIPLE. Erasing, or resetting, a bit of information costs at least a szilard of work.

LAWS OF THERMODYNAMICS. Backbones of the theory of thermodynamics.

- **ZEROTH LAW OF THERMODYNAMICS.** Establishes the notion of a thermometer. If a spoon of Baxter's is in thermal equilibrium with an almond pudding of Audrey's and with a curry of Caspian's, then Audrey's almond pudding is in thermal equilibrium with Caspian's curry.
- **FIRST LAW OF THERMODYNAMICS.** The energy of every closed, isolated system is conserved.
- **SECOND LAW OF THERMODYNAMICS.** Star in the cast of thermodynamics. One formulation is, the entropy of every closed, isolated system can only increase or remain constant.
- **THIRD LAW OF THERMODYNAMICS.** No process (of finitely many steps) can cool a system to absolute-zero temperature.

LINEAR ALGEBRA. Branch of mathematics that underlies quantum computing. Requires that one solve loads of equations simultaneously. If you ever want to insult a quantum information theorist, say, "Pfft. Isn't quantum information just linear algebra?"

LIVER. See *entropy*.

LOCALIZATION. Confinement to a finite volume, in contrast with the spread-out nature of a wave that represents a quantum system. For a slightly different usage, see *many-body localization*.

LOGARITHM. Function that converts products of numbers into sums of numbers.

MACROSTATE. Consists of the large-scale properties (such as the temperature and pressure) of a collection of particles.

MANY-BODY LOCALIZATION. Phase of quantum matter in which particles mostly stay put and entanglement spreads slowly. Contrasted with the thermal phase.

MASER. Like a laser, but emits microwave radiation instead of classical light.

MATRIX. Collection of numbers arranged in a grid. Mathematical representation of a quantum observable. Not all matrices commute with each other. See *übernumber*.

MAXIMAL ENTANGLEMENT. Relationship that quantum particles can share.

Measurements of maximally entangled states can be correlated as strongly as anything in our world can be (as far as we know).

MAXWELL'S DEMON. "Finite being" dreamed up by nineteenth-century physicist James Clerk Maxwell. Star of a thermodynamic paradox. Appears, prima facie, to violate the second law of thermodynamics.

MBL-MOBILE. Quantum engine whose operation involves transitioning quantum particles between a many-body-localized phase and a thermal phase.

MEASUREMENT DISTURBANCE. Unavoidable alteration of a quantum system's state by a measurement of the system.

METROLOGY. Study of how, and how well, we can measure things.

MICROCANONICAL ENSEMBLE. What Napoleon would have called part of his army. Also, the classical analog of the microcanonical quantum state.

MICROCANONICAL STATE. Equilibrium state of a closed, isolated quantum system.

MICROSTATE. Property of a collection of particles. Defined by a list of the particles' positions and momenta (and, depending on what the particles consist of, their angular momenta, their vibrations, etc.).

MOLECULAR SWITCH. Molecule that can change shape from one configuration to another. Found in natural systems, such as our eyes, and used in technologies, such as solar-fuel-storage devices.

MOMENTUM. Property that reflects the difficulty of stopping an object. The heavier the object, and the more quickly it moves, the greater its momentum.

MONOGAMY OF ENTANGLEMENT. Limitation on the amount of entanglement that one particle can share with others. The more entanglement that a particle of Baxter's shares with a particle of Audrey's, the less entanglement Baxter's particle can share with a particle of Caspian's.

NEGATIVE TEMPERATURE. Temperature below absolute zero, achievable by systems whose energies are quantized. A system is hotter at all negative temperatures than it would be at any positive temperature.

NONCLASSICAL. Inaccurately described by classical mechanics, electrodynamics, and general relativity. Quantum physics is nonclassical.

NONEQUILIBRIUM THERMODYNAMICS. Study of the energy of systems away from equilibrium, or roiled up.

NUCLEAR MAGNETIC RESONANCE (NMR). Experimental toolkit used to control nuclei that store quantum information in quantum computers, as well as to image brains in magnetic resonance imaging (MRI).

OBSERVABLE. Measurable property. Examples include position and momentum.

ONE-SHOT INFORMATION THEORY. Study of how efficiently we can perform information-processing tasks (solve computational problems, communicate information, secure information, and store information) in a limited number of trials, or given a limited amount of information.

ONE-SHOT THERMODYNAMICS. Study of how adroitly we can perform thermodynamic tasks (such as work extraction) in a limited number of trials, or given small systems.

OPEN SYSTEM. System that interacts with other systems, such as by exchanging energy with a heat bath.

OPERATIONAL. Concerned with how efficiently agents can perform tasks with given resources—for example, how efficiently one can transmit information, given a staticky telephone, or can extract work, given heat baths at different temperatures. Information theory and thermodynamics are operational.

PAULI'S EXCLUSION PRINCIPLE. No two fermions can be in the same quantum state. Explains how electrons arrange themselves in atoms.

PERPETUUM MOBILE. Perpetual-motion machine. Forbidden by the second law of thermodynamics.

PHASE. Form in which matter can exist. Everyday examples include solid, liquid, and gas. Quantum examples include many-body localization.

PHOTODETECTOR. Camera that collects light, registering photons.

PHOTOISOMER. Molecular switch.

PHOTON. Particle of light.

PLATFORM. Type of hardware—for example, a material used to build a quantum computer.

POWER. The rate at which an object delivers energy, e.g., at which an engine performs work.

PRIME NUMBER. Number that's divisible only by itself and by one.

PRINCIPLE OF NO SIGNALING. Information can't travel more quickly than light. Originates in Einstein's theory of relativity and obeyed by entangled systems.

PROBABILISTIC. Random. Contrasted with *deterministic*.

PROBABILITY DISTRIBUTION. Set of probabilities that describe how a random event might unfold.

QUANTIZATION. Limitation to only a few possible numbers. Examples of quantization include the energy that a hydrogen atom has due to its one electron's motion and the electron's attraction to the nucleus.

QUANTUM. Indivisible unit, as of energy or light.

QUANTUM ADIABATICITY. Very slow adjustment of a quantum system's energy ladder, as by strengthening an electric field near an atom. The system remains on the same ladder rung throughout the adjustment.

QUANTUM COMPUTER. Computer whose operation relies on quantum phenomena. Able to solve certain computational problems, such as the factoring of prime numbers, far more quickly than any classical computer.

QUANTUM DOT. Artificial atom—in some cases, a little patch of space on a semiconductor surface. An electron is confined to the patch, by an electric field, similarly to how an electron in an atom is confined to remain near the nucleus. Able to store one unit of quantum information.

QUANTUM INFORMATION. Information that can be stored in, and processed by, quantum systems.

QUANTUM INFORMATION SCIENCE. Study of how we can use quantum resources (such as entanglement) to process information in ways impossible with only classical resources.

QUANTUM-INFORMATION THERMODYNAMICS. Intersection of quantum computing and thermodynamics.

QUANTUM SPEEDUP. Outperformance of a classical system by a quantum system on an information-processing task.

QUANTUM STATE. Mathematical representation of a quantum system's status. Usable to predict the probability that a given measurement will yield a given outcome. Quantum analog of a probability distribution.

QUANTUM STEAMPUNK. Reenvisioning of nineteenth-century thermodynamics for small, quantum, far-from-equilibrium, and information-processing systems. Intersection of quantum computing and thermodynamics, plus the use of this intersection as a new lens onto other disciplines. Shares its aesthetic with steampunk in juxtaposing futuristic technology (quantum computing) with a Victorian setting (thermodynamics).

QUANTUM THEORY. Study of small systems (such as electrons, protons, and photons) that fall outside the purview of classical physics.

QUANTUM THERMODYNAMICS. Extension of conventional thermodynamics to quantum systems.

QUANTUM THERMOMETRY. Study of the effects of quantum phenomena on temperature measurements.

QUBIT. Basic unit of quantum information.

REAL NUMBER. Number of the sort that describes our everyday lives. A negative number, a positive number, or zero.

RESOURCE. Something that's scarce and that's valuable because it's useful.

RESOURCE THEORY. Simple model, developed in quantum information theory, for any situation in which constraints restrict the systems that one can access and the operations that one can perform.

RESOURCE-THEORY FRAMEWORK. Mathematical and conceptual toolkit of resource theories.

ROBIN HOOD TRANSFER. Transformation of one probability distribution into another, akin to the transformation of one distribution of wealth into another via a theft from the rich and a gift to the poor.

SCRAMBLING. Dissemination of initially localized quantum information across a system through many-particle entanglement.

SCRAMBLING SIGNAL. Number that indicates whether a quantum system is scrambled.

"SECOND LAWS" OF THERMODYNAMICS. Equations or inequalities that are stronger—that provide more information—than the second law of thermodynamics.

SHORTCUT TO ADIABATICITY. Quick adjustment of a quantum system's energy ladder. After the adjustment, the system ideally ends on the same ladder rung on which it began.

SIMULATOR. Special-purpose computer that calculates how a certain system would behave under certain conditions—for example, how a certain material would respond if heated.

SPIN. Property that quantum systems have and that classical systems lack. Described by the same mathematics as angular momentum.

SQUEEZED LIGHT. Light whose quantum uncertainty is squeezed into one observable, leaving another observable with a nearly well-defined value.

STATISTICAL MECHANICS. Study of many-particle systems. Less operational than thermodynamics.

STEAMPUNK. Genre of literature, art, and film in which Victorian-era settings are juxtaposed with futuristic technologies.

SUPERCONDUCTING QUBIT. Tiny circuit cooled to a low temperature and able to store a unit of quantum information.

SUPERCONDUCTIVITY. Property that graces certain quantum materials. Current can flow through the material forever, without dissipating.

SUPERPOSITION. Sum of waves that is a wave itself.

SURPRISAL. How much information you gain upon learning how a random event unfolded.

SZILARD. Maximum amount of energy obtainable from one run of Szilard's engine. Minimum amount of energy required to erase a bit of information. Named after a twentieth-century Hungarian-American physicist.

SZILARD'S ENGINE. Engine usable to turn heat into work, with help from a bit of information.

THERMAL EQUILIBRIUM. One system is in thermal equilibrium with another if both have the same temperature.

THERMAL MACHINE. Device that uses, produces, or stores heat or work. Examples include heat engines, refrigerators, heat pumps, ratchets, batteries, and clocks.

THERMAL PHASE. Phase of quantum matter in which particles move, and entanglement spreads, quickly. Contrasted with the many-body localized phase.

THERMODYNAMIC LIMIT. Idealization, focused on in traditional thermodynamics, in which systems are infinitely large.

THERMODYNAMICS. Study of energy—the forms it can assume and its transformations among those forms. Colored by operationalism.

THERMODYNAMIC UNCERTAINTY RELATION. Inequality that interrelates the entropy produced when particles flow from one bath to another with the fluctuations in the current of particles.

TRANS CONFIGURATION. "Open" configuration of a photoisomer, or molecular switch.

ÜBERNUMBER. Collection of numbers that, in some ways, acts like a number itself. See *matrix*.

UNCERTAINTY PRINCIPLE. Limitation on the extent to which one quantum observable (such as position) has a well-defined value, given the extent to which another observable (such as momentum) has a well-defined value.

UNCERTAINTY RELATION, QUANTUM. Inequality that encapsulates the uncertainty principle.

UNIVERSAL COMPUTER. Computer that can be programmed to solve any solvable computational problem, then reprogrammed to solve any other.

VELOCITY. Speed at which, and direction in which, an object is moving.

WAVE FUNCTION. Certain mathematical representation of a quantum state. Emphasizes the state's wavelike properties.

WAVE-FUNCTION COLLAPSE. See *measurement disturbance*.

WAVELENGTH. Distance between two consecutive crests of a wave.

WAVE-PARTICLE DUALITY. Every chunk of matter and light resembles a wave in some ways and resembles a particle in others.

WORK. Coordinated, organized energy that's being transferred between systems and that can be directly harnessed to perform a useful task, such as pushing a car up a hill.

WORK EXTRACTION. Acquisition of useful energy, as from two different-temperature heat baths via an engine.

REFERENCES

CHAPTER 0 —·— PROLOGUE

1. Malik, Wajeeha. "Inky's Daring Escape Shows How Smart Octopuses Are." *National Geographic*, April 14, 2016. https://www.nationalgeographic.com/animals/article/160414-inky-octopus-escapes-intelligence.

CHAPTER 1 —·— INFORMATION THEORY

1. Schumacher, Benjamin, and Michael Westmoreland. *Quantum Processes, Systems, and Information*. New York: Cambridge University Press, 2010.
2. Munroe, Randall. *What If? Serious Scientific Answers to Absurd Hypothetical Questions*. International ed. Boston: Mariner Books, 2014.
3. Suzuki, Jeff. *A History of Mathematics*. Upper Saddle River, NJ: Prentice Hall, 2002.
4. "Liver: Anatomy and Functions." Johns Hopkins Medicine. Accessed April 4, 2021. https://www.hopkinsmedicine.org/health/conditions-and-diseases/liver-anatomy-and-functions.
5. Tribus, M., and E. C. McIrvine. "Energy and Information." *Scientific American* 225, no. 3 (September 1971): 179–88, quote at p. 180. http://www.esalq.usp.br/lepse/imgs/conteudo_thumb/Energy-and-Information.pdf.

CHAPTER 2 —·— QUANTUM PHYSICS

1. Improbable Research (blog). "Yet Another Prize for Ig-Winning Ponytail-Physics Researcher," December 15, 2015. https://www.improbable.com/2015/12/15/yet-another-prize-for-ig-winning-ponytail-physics-researcher/.
2. Sebens, Charles T. "How Electrons Spin." *Studies in History and Philosophy of Science Part B: Studies in History and Philosophy of Modern Physics* 68 (November 1, 2019): 40–50. https://doi.org/10.1016/j.shpsb.2019.04.007.
3. Heisenberg, W. "Über den anschaulichen Inhalt der quantentheoretischen Kinematik und Mechanik." *Zeitschrift für Physik* 43, no. 3 (March 1, 1927): 172–98. https://doi.org/10.1007/BF01397280.

4. Kennard, E. H. "Zur Quantenmechanik einfacher Bewegungstypen." *Zeitschrift für Physik* 44, no. 4 (April 1, 1927): 326–52. https://doi.org/10.1007/BF01391200.
5. Bell, J. S. "On the Einstein Podolsky Rosen Paradox." *Physics Physique Fizika* 1, no. 3 (November 1, 1964): 195–200. https://doi.org/10.1103/PhysicsPhysiqueFizika.1.195.
6. John Gribbin. *Schrödinger's Kittens and the Search for Reality.* New York: Back Bay Books, 1995.

CHAPTER 3 —·— QUANTUM COMPUTATION

1. Feynman, Richard P. "Simulating Physics with Computers." *International Journal of Theoretical Physics* 21, no. 6 (June 1, 1982): 467–88. https://doi.org/10.1007/BF02650179.
2. Manin, Yuri. *Computable and Uncomputable*. Moscow: Sovetskoye Radio, 1980.
3. Benioff, Paul. "The Computer as a Physical System: A Microscopic Quantum Mechanical Hamiltonian Model of Computers as Represented by Turing Machines." *Journal of Statistical Physics* 22, no. 5 (May 1, 1980): 563–91. https://doi.org/10.1007/BF01011339.
4. Fredkin, Edward, and Tommaso Toffoli. "Conservative Logic." *International Journal of Theoretical Physics* 21, no. 3 (April 1, 1982): 219–53. https://doi.org/10.1007/BF01857727.
5. Deutsch, David. "Quantum Theory, the Church–Turing Principle and the Universal Quantum Computer." *Proceedings of the Royal Society of London A. Mathematical and Physical Sciences* 400, no. 1818 (July 8, 1985): 97–117. https://doi.org/10.1098/rspa.1985.0070.
6. Altman, Ehud, Kenneth R. Brown, Giuseppe Carleo, Lincoln D. Carr, Eugene Demler, Cheng Chin, et al. "Quantum Simulators: Architectures and Opportunities." *PRX Quantum* 2, no. 1 (February 24, 2021): 017003. https://doi.org/10.1103/PRXQuantum.2.017003.

CHAPTER 4 —·— THERMODYNAMICS

1. Grahame, Kenneth. *The Wind in the Willows*. New York: Charles Scribner's Sons, 1913. https://www.gutenberg.org/files/27805/27805-h/27805-h.htm.
2. Prigogine, Ilya. "Nobel Lecture: Time Structure and Fluctuations." Nobel

Prize website, "The Nobel Prize in Chemistry 1977." Accessed April 4, 2021. https://www.nobelprize.org/prizes/chemistry/1977/prigogine/lecture/.
3. Prigogine, Ilya. "Biographical." In *Nobel Lectures, Chemistry 1971-1980*, translated from the French, edited by Tore Frängsmyr and Sture Forsén, Singapore: World Scientific Publishing, 1993. https://www.nobelprize.org/prizes/chemistry/1977/prigogine/biographical/.
4. Fowler, R. H., and E. A. Guggenheim. *Statistical Thermodynamics: A Version of Statistical Mechanics for Students of Physics and Chemistry*. New York: Macmillan; Cambridge, UK: Cambridge University Press, 1939.
5. Fernández-Pineda, C., and S. Velasco "Comment on 'Historical Observations on Laws of Thermodynamics.'" *Journal of Chemical & Engineering Data* 57, no. 4 (April 12, 2012): 1347–1347. https://doi.org/10.1021/je300082q.
6. Eddington, Arthur. *The Nature of the Physical World*. 1928. Reprint, Cambridge, UK: Cambridge University Press, 2007. https://henry.pha.jhu.edu/Eddington.2008.pdf.
7. Lloyd, Seth. "Going into Reverse." *Nature* 430, no. 7003 (August 2004): 971–971. https://doi.org/10.1038/430971a.
8. Son, Hyungmok, Juliana J. Park, Wolfgang Ketterle, and Alan O. Jamison. "Collisional Cooling of Ultracold Molecules." *Nature* 580, no. 7802 (April 2020): 197–200. https://doi.org/10.1038/s41586-020-2141-z.

CHAPTER 5 —·— A FINE MERGER

1. Szilard, Leo. "On the Decrease of Entropy in a Thermodynamic System by the Intervention of Intelligent Beings." *Behavioral Science* 9, no. 4 (1964): 301–10. https://doi.org/10.1002/bs.3830090402.
2. Landauer, R. "Irreversibility and Heat Generation in the Computing Process." *IBM Journal of Research and Development* 5, no. 3 (July 1961): 183–91. https://doi.org/10.1147/rd.53.0183.
3. Bennett, Charles H. "Demons, Engines and the Second Law." *Scientific American* 257, no. 5 (November 1987): 108–116. https://www.jstor.org/stable/24979551.
4. Bender, Carl M., Dorje C. Brody, and Bernhard J. Meister. "Unusual Quantum States: Non–Locality, Entropy, Maxwell's Demon and Fractals." *Proceedings of the Royal Society A: Mathematical, Physical and Engineering Sciences* 461, no. 2055 (March 8, 2005): 733–53. https://doi.org/10.1098/rspa.2004.1351.
5. Rio, Lídia del, Johan Åberg, Renato Renner, Oscar Dahlsten, and Vlatko Vedral. "The Thermodynamic Meaning of Negative Entropy." *Nature* 476, no. 7361 (August 2011): 476–476. https://doi.org/10.1038/nature10395.

6. Kim, Sang Wook, Takahiro Sagawa, Simone De Liberato, and Masahito Ueda. "Quantum Szilard Engine." *Physical Review Letters* 106, no. 7 (February 14, 2011): 070401. https://doi.org/10.1103/PhysRevLett.106.070401.
7. Szilard, Leo. "On the Decrease of Entropy in a Thermodynamic System by the Intervention of Intelligent Beings." *Behavioral Science* 9, no. 4 (1964): 301–10. https://doi.org/https://doi.org/10.1002/bs.3830090402.
8. Bennett, Charles H. "The Thermodynamics of Computation—A Review." *International Journal of Theoretical Physics* 21, no. 12 (December 1, 1982): 905–40. https://doi.org/10.1007/BF02084158.

CHAPTER 6 —·— THE PHYSICS OF YESTERDAY'S TOMORROW

1. Watanabe, Satoshi, and Louis de Broglie. *Le Deuxième Théorème de La Thermodynamique et La Mécanique Ondulatoire.* Hermann, 1935.
2. Slater, J. C. *Introduction to Chemical Physics.* 1st ed. New York: McGraw-Hill, 1939, 46.
3. Demers, Pierre. "Le Second Principe et La Théorie Des Quanta." *Canadian Journal of Research* 11, no. 50 (1944): 27–51.
4. Ramsey, Norman F. "Thermodynamics and Statistical Mechanics at Negative Absolute Temperatures." *Physical Review* 103, no. 1 (July 1, 1956): 20–28. https://doi.org/10.1103/PhysRev.103.20.
5. Scovil, H. E. D., and E. O. Schulz-DuBois. "Three-Level Masers as Heat Engines." *Physical Review Letters* 2, no. 6 (March 15, 1959): 262–63. https://doi.org/10.1103/PhysRevLett.2.262.
6. Geusic, J. E., E. O. Schulz-DuBios, and H. E. D. Scovil. "Quantum Equivalent of the Carnot Cycle." *Physical Review* 156, no. 2 (April 10, 1967): 343–51. https://doi.org/10.1103/PhysRev.156.343.
7. Lindblad, G. "On the Generators of Quantum Dynamical Semigroups." *Communications in Mathematical Physics* 48, no. 2 (June 1, 1976): 119–30. https://doi.org/10.1007/BF01608499.
8. Gorini, Vittorio, Andrzej Kossakowski, and E. C. G. Sudarshan. "Completely Positive Dynamical Semigroups of N-level Systems." *Journal of Mathematical Physics* 17, no. 5 (May 1, 1976): 821–25. https://doi.org/10.1063/1.522979.
9. Park, James L., and William Band. "Generalized Two-Level Quantum Dynamics. III. Irreversible Conservative Motion." *Foundations of Physics* 8, no. 3 (April 1, 1978): 239–54. https://doi.org/10.1007/BF00715210.
10. Kraus, K. "General State Changes in Quantum Theory." *Annals of Physics* 64, no. 2 (June 1, 1971): 311–35. https://doi.org/10.1016/0003-4916(71)90108-4.

11. Davies, E. B. "Markovian Master Equations." *Communications in Mathematical Physics* 39, no. 2 (June 1, 1974): 91–110. https://doi.org/10.1007/BF01608389.
12. Kosloff, Ronnie. "A Quantum Mechanical Open System as a Model of a Heat Engine." *Journal of Chemical Physics* 80, no. 4 (February 15, 1984): 1625–31. https://doi.org/10.1063/1.446862.
13. Alicki, Robert. "The Quantum Open System as a Model of the Heat Engine." *Journal of Physics A: Mathematical and General* 12, no. 5 (1979): L103-07. https://iopscience.iop.org/article/10.1088/0305-4470/12/5/007.
14. Scully, Robert J., and Marlan O. Scully. *The Demon and the Quantum: From the Pythagorean Mystics to Maxwell's Demon and Quantum Mystery*. 2nd ed. Weinheim, Germany: Wiley-VCH, 2010.
15. Lloyd, Seth. "Black Holes, Demons, and the Loss of Coherence: How Complex Systems Get Information, and What They Do with It." PhD diss., Rockefeller University, 1988.
16. Goldstein, Sheldon, Joel L. Lebowitz, Roderich Tumulka, and Nino Zanghì. "Canonical Typicality." *Physical Review Letters* 96, no. 5 (February 8, 2006): 050403. https://doi.org/10.1103/PhysRevLett.96.050403.
17. Popescu, Sandu, Anthony J. Short, and Andreas Winter. "Entanglement and the Foundations of Statistical Mechanics." *Nature Physics* 2, no. 11 (November 2006): 754–58. https://doi.org/10.1038/nphys444.
18. Page, Don N. "Black Hole Information." *ArXiv:Hep-Th/9305040*, February 25, 1995. http://arxiv.org/abs/hep-th/9305040.
19. Prigogine, I., and C. George. "The Second Law as a Selection Principle: The Microscopic Theory of Dissipative Processes in Quantum Systems." *Proceedings of the National Academy of Sciences* 80, no. 14 (July 1, 1983): 4590–94. https://doi.org/10.1073/pnas.80.14.4590.
20. Anderson, P. W. "More Is Different." *Science* 177, no. 4047 (August 4, 1972): 393–96. https://doi.org/10.1126/science.177.4047.393.
21. Frenzel, Max F., David Jennings, and Terry Rudolph. "Reexamination of Pure Qubit Work Extraction." *Physical Review E* 90, no. 5 (November 18, 2014): 052136. https://doi.org/10.1103/PhysRevE.90.052136.

CHAPTER 7 —·— PEDAL TO THE METAL

1. Scovil, H. E. D., and E. O. Schulz-DuBois. "Three-Level Masers as Heat Engines." *Physical Review Letters* 2, no. 6 (March 15, 1959): 262–63. https://doi.org/10.1103/PhysRevLett.2.262.
2. Geusic, J. E., E. O. Schulz-DuBios, and H. E. D. Scovil. "Quantum Equivalent

of the Carnot Cycle." *Physical Review* 156, no. 2 (April 10, 1967): 343–51. https://doi.org/10.1103/PhysRev.156.343.

3. Kalaee, Alex Arash Sand, Andreas Wacker, and Patrick P. Potts. "Violating the Thermodynamic Uncertainty Relation in the Three-Level Maser." *ArXiv:2103.07791 [Quant-Ph]*, March 13, 2021. http://arxiv.org/abs/2103.07791.
4. Campisi, Michele, and Rosario Fazio. "The Power of a Critical Heat Engine." *Nature Communications* 7, no. 1 (June 20, 2016): 11895. https://doi.org/10.1038/ncomms11895.
5. Oz-Vogt, J., A. Mann, and M. Revzen. "Thermal Coherent States and Thermal Squeezed States." *Journal of Modern Optics* 38, no. 12 (December 1, 1991): 2339–47. https://doi.org/10.1080/09500349114552501.
6. Roßnagel, J., O. Abah, F. Schmidt-Kaler, K. Singer, and E. Lutz. "Nanoscale Heat Engine beyond the Carnot Limit." *Physical Review Letters* 112, no. 3 (January 22, 2014): 030602. https://doi.org/10.1103/PhysRevLett.112.030602.
7. Niedenzu, Wolfgang, David Gelbwaser-Klimovsky, Abraham G. Kofman, and Gershon Kurizki. "On the Operation of Machines Powered by Quantum Non-Thermal Baths." *New Journal of Physics* 18, no. 8 (August 2, 2016): 083012. https://doi.org/10.1088/1367-2630/18/8/083012.
8. Gardas, Bartłomiej, and Sebastian Deffner. "Thermodynamic Universality of Quantum Carnot Engines." *Physical Review E* 92, no. 4 (October 12, 2015): 042126. https://doi.org/10.1103/PhysRevE.92.042126.
9. Klaers, Jan, Stefan Faelt, Atac Imamoglu, and Emre Togan. "Squeezed Thermal Reservoirs as a Resource for a Nanomechanical Engine beyond the Carnot Limit." *Physical Review* X 7, no. 3 (September 13, 2017): 031044. https://doi.org/10.1103/PhysRevX.7.031044.
10. Yunger Halpern, Nicole, Christopher David White, Sarang Gopalakrishnan, and Gil Refael. "Quantum Engine Based on Many-Body Localization." *Physical Review B* 99, no. 2 (January 22, 2019): 024203. https://doi.org/10.1103/PhysRevB.99.024203.
11. Palao, José P., Ronnie Kosloff, and Jeffrey M. Gordon. "Quantum Thermodynamic Cooling Cycle." *Physical Review E* 64, no. 5 (October 30, 2001): 056130. https://doi.org/10.1103/PhysRevE.64.056130.
12. Linden, Noah, Sandu Popescu, and Paul Skrzypczyk. "How Small Can Thermal Machines Be? The Smallest Possible Refrigerator." *Physical Review Letters* 105, no. 13 (September 21, 2010): 130401. https://doi.org/10.1103/PhysRevLett.105.130401.
13. Binder, Felix C., Sai Vinjanampathy, Kavan Modi, and John Goold. "Quantacell: Powerful Charging of Quantum Batteries." *New Journal of Physics*

17, no. 7 (July 22, 2015): 075015. https://doi.org/10.1088/1367-2630/17/7/075015.

14. Maslennikov, Gleb, Shiqian Ding, Roland Hablützel, Jaren Gan, Alexandre Roulet, Stefan Nimmrichter et al. "Quantum Absorption Refrigerator with Trapped Ions." *Nature Communications* 10, no. 1 (January 14, 2019): 202. https://doi.org/10.1038/s41467-018-08090-0.

CHAPTER 8 —·— TICK TOCK

1. Brewer, S. M., J.-S. Chen, A. M. Hankin, E. R. Clements, C. W. Chou, D. J. Wineland, D. B. Hume, and D. R. Leibrandt. "$^{27}Al^{+}$ Quantum-Logic Clock with a Systematic Uncertainty below 10^{-18}." *Physical Review Letters* 123, no. 3 (July 15, 2019): 033201. https://doi.org/10.1103/PhysRevLett.123.033201.
2. Dubé, Pierre. "Ion Clock Busts into New Precision Regime." *Physics* 12 (July 15, 2019). https://physics.aps.org/articles/v12/79.
3. Newell, David B., and Eite Tiesinga. "Reference on Constants, Units, and Uncertainty: International System of Units (SI)." NIST website. Accessed May 6, 2021. https://physics.nist.gov/cuu/Units/current.html.
4. Pauli, Wolfgang. *Handbuch Der Physik*. 1st ed. Vol. 23. Berlin: Springer, 1926.
5. Pauli, Wolfgang. *Handbuch Der Physik*. 2nd ed. Vol. 24. Berlin: Springer, 1933.
6. Pauli, Wolfgang. *Handbuch Der Physik*. Vol. 5, Part 1: Prinzipien der Quantentheorie I. Berlin: Springer, 1958.
7. Woods, Mischa P., Ralph Silva, and Jonathan Oppenheim. "Autonomous Quantum Machines and Finite-Sized Clocks." *Annales Henri Poincaré* 20, no. 1 (January 1, 2019): 125–218. https://doi.org/10.1007/s00023-018-0736-9.
8. Yunger Halpern, Nicole, and David T. Limmer. "Fundamental Limitations on Photoisomerization from Thermodynamic Resource Theories." *Physical Review A* 101, no. 4 (April 17, 2020): 042116. https://doi.org/10.1103/PhysRevA.101.042116

CHAPTER 9 —·— UNSTEADY AS SHE GOES

1. "Selling the Victorians." The National Archive website. Accessed April 5, 2021. https://www.nationalarchives.gov.uk/education/resources/selling-the-victorians/.
2. Mossa, A., M. Manosas, N. Forns, J. M. Huguet, and F. Ritort. "Dynamic Force Spectroscopy of DNA Hairpins: I. Force Kinetics and Free Energy Landscapes." *Journal of Statistical Mechanics: Theory and Experiment* 2009, no. 2 (February 25, 2009): P02060. https://doi.org/10.1088/1742-5468/2009/02/P02060.
3. Schroeder, Daniel V. *An Introduction to Thermal Physics*. San Francisco: Pearson, 1999.

4. Jarzynski, C. "Nonequilibrium Equality for Free Energy Differences." *Physical Review Letters* 78, no. 14 (April 7, 1997): 2690–93. https://doi.org/10.1103/PhysRevLett.78.2690.
5. Crooks, Gavin E. "Entropy Production Fluctuation Theorem and the Nonequilibrium Work Relation for Free Energy Differences." *Physical Review E* 60, no. 3 (September 1, 1999): 2721–26. https://doi.org/10.1103/PhysRevE.60.2721.
6. Liphardt, Jan, Sophie Dumont, Steven B. Smith, Ignacio Tinoco, and Carlos Bustamante. "Equilibrium Information from Nonequilibrium Measurements in an Experimental Test of Jarzynski's Equality." *Science* 296, no. 5574 (June 7, 2002): 1832–35. https://doi.org/10.1126/science.1071152.
7. Hummer, Gerhard, and Attila Szabo. "Free Energy Reconstruction from Nonequilibrium Single-Molecule Pulling Experiments." *Proceedings of the National Academy of Sciences* 98, no. 7 (March 27, 2001): 3658–61. https://doi.org/10.1073/pnas.071034098.
8. Blickle, V., T. Speck, L. Helden, U. Seifert, and C. Bechinger. "Thermodynamics of a Colloidal Particle in a Time-Dependent Nonharmonic Potential." *Physical Review Letters* 96, no. 7 (February 23, 2006): 070603. https://doi.org/10.1103/PhysRevLett.96.070603.
9. Douarche, F., S. Ciliberto, A. Petrosyan, and I. Rabbiosi. "An Experimental Test of the Jarzynski Equality in a Mechanical Experiment." *EPL (Europhysics Letters)* 70, no. 5 (April 29, 2005): 593. https://doi.org/10.1209/epl/i2005-10024-4.
10. Misof, K., W. J. Landis, K. Klaushofer, and P. Fratzl. "Collagen from the Osteogenesis Imperfecta Mouse Model (OIM) Shows Reduced Resistance against Tensile Stress." *Journal of Clinical Investigation* 100, no. 1 (July 1, 1997): 40–45. https://doi.org/10.1172/JCI119519.
11. Herczenik, Eszter, and Martijn F. B. G. Gebbink. "Molecular and Cellular Aspects of Protein Misfolding and Disease." *FASEB Journal* 22, no. 7 (2008): 2115–33. https://doi.org/https://doi.org/10.1096/fj.07-099671.
12. Utsumi, Y., D. S. Golubev, M. Marthaler, K. Saito, T. Fujisawa, and Gerd Schön. "Bidirectional Single-Electron Counting and the Fluctuation Theorem." *Physical Review B* 81, no. 12 (March 29, 2010): 125331. https://doi.org/10.1103/PhysRevB.81.125331.
13. Küng, B., C. Rössler, M. Beck, M. Marthaler, D. S. Golubev, Y. Utsumi et al. "Irreversibility on the Level of Single-Electron Tunneling." *Physical Review X* 2, no. 1 (January 13, 2012): 011001. https://doi.org/10.1103/PhysRevX.2.011001.
14. Saira, O.-P., Y. Yoon, T. Tanttu, M. Möttönen, D. V. Averin, and J. P. Pekola.

"Test of the Jarzynski and Crooks Fluctuation Relations in an Electronic System." *Physical Review Letters* 109, no. 18 (October 31, 2012): 180601. https://doi.org/10.1103/PhysRevLett.109.180601.

15. Bartolotta, Anthony, and Sebastian Deffner. "Jarzynski Equality for Driven Quantum Field Theories." *Physical Review* X 8, no. 1 (February 27, 2018): 011033. https://doi.org/10.1103/PhysRevX.8.011033.
16. Ortega, Alvaro, Emma McKay, Álvaro M. Alhambra, and Eduardo Martín-Martínez. "Work Distributions on Quantum Fields." *Physical Review Letters* 122, no. 24 (June 21, 2019): 240604. https://doi.org/10.1103/PhysRevLett.122.240604.
17. Bruschi, David, Benjamin Morris, and Ivette Fuentes. "Thermodynamics of Relativistic Quantum Fields Confined in Cavities." *Physics Letters A* 384, no. 25 (September 7, 2020): 126601. https://doi.org/10.1016/j.physleta.2020.126601.
18. Teixidó-Bonfill, Adam, Alvaro Ortega, and Eduardo Martín-Martínez. "First Law of Quantum Field Thermodynamics." *Physical Review A* 102, no. 5 (November 18, 2020): 052219. https://doi.org/10.1103/PhysRevA.102.052219.
19. Liu, Nana, John Goold, Ivette Fuentes, Vlatko Vedral, Kavan Modi, and David Bruschi. "Quantum Thermodynamics for a Model of an Expanding Universe." *Classical and Quantum Gravity* 33, no. 3 (January 11, 2016): 035003.
20. An, Shuoming, Jing-Ning Zhang, Mark Um, Dingshun Lv, Yao Lu, Junhua Zhang et al. "Experimental Test of the Quantum Jarzynski Equality with a Trapped-Ion System." *Nature Physics* 11, no. 2 (February 2015): 193–99. https://doi.org/10.1038/nphys3197.
21. Batalhão, Tiago B., Alexandre M. Souza, Laura Mazzola, Ruben Auccaise, Roberto S. Sarthour, Ivan S. Oliveira et al. "Experimental Reconstruction of Work Distribution and Study of Fluctuation Relations in a Closed Quantum System." *Physical Review Letters* 113, no. 14 (October 3, 2014): 140601. https://doi.org/10.1103/PhysRevLett.113.140601.
22. Naghiloo, M., J. J. Alonso, A. Romito, E. Lutz, and K. W. Murch. "Information Gain and Loss for a Quantum Maxwell's Demon." *Physical Review Letters* 121, no. 3 (July 17, 2018): 030604. https://doi.org/10.1103/PhysRevLett.121.030604.
23. Zhang, Zhenxing, Tenghui Wang, Liang Xiang, Zhilong Jia, Peng Duan, Weizhou Cai et al. "Experimental Demonstration of Work Fluctuations along a Shortcut to Adiabaticity with a Superconducting Xmon Qubit." *New Journal of Physics* 20, no. 8 (August 2, 2018): 085001. https://doi.org/10.1088/1367-2630/aad4e7.
24. Cerisola, Federico, Yair Margalit, Shimon Machluf, Augusto J. Roncaglia, Juan Pablo Paz, and Ron Folman. "Using a Quantum Work Meter to Test

Non-Equilibrium Fluctuation Theorems." *Nature Communications* 8, no. 1 (November 1, 2017): 1241. https://doi.org/10.1038/s41467-017-01308-7.
25. Hernández-Gómez, S., S. Gherardini, F. Poggiali, F. S. Cataliotti, A. Trombettoni, P. Cappellaro et al. "Experimental Test of Exchange Fluctuation Relations in an Open Quantum System." *Physical Review Research* 2, no. 2 (June 12, 2020): 023327. https://doi.org/10.1103/PhysRevResearch.2.023327.

CHAPTER 10 —·— ENTROPY, ENERGY, AND A TINY POSSIBILITY

1. Rényi, Alfréd. "On Measures of Entropy and Information," *Proceedings of the Fourth Berkeley Symposium on Mathematical Statistics and Probability*, Vol. 1, 547–61. University of California, Berkeley: University of California Press, 1961.
2. Faist, Philippe. "Welcome to the Entropy Zoo." Personal website of Philippe Faist. Accessed April 5, 2021. https://phfaist.com/d/entropyzoo/TheEntropyZoo.pdf.
3. Rio, Lídia del, Johan Åberg, Renato Renner, Oscar Dahlsten, and Vlatko Vedral. "The Thermodynamic Meaning of Negative Entropy." *Nature* 476, no. 7361 (August 2011): 476–476. https://doi.org/10.1038/nature10395.
4. Yunger Halpern, Nicole, Andrew J. P. Garner, Oscar C. O. Dahlsten, and Vlatko Vedral. "Introducing One-Shot Work into Fluctuation Relations." *New Journal of Physics* 17, no. 9 (September 11, 2015): 095003. https://doi.org/10.1088/1367-2630/17/9/095003.
5. Burnette, Joyce. "Women Workers in the British Industrial Revolution." EH.Net Encyclopedia, edited by Robert Whaples, March 26, 2008. https://eh.net/encyclopedia/women-workers-in-the-british-industrial-revolution/.
6. Lamb, Evelyn. "5 Sigma: What's That?" *Scientific American* (blog), July 17, 2012. https://blogs.scientificamerican.com/observations/five-sigmawhats-that/.
7. Jarzynski, Christopher. "Rare Events and the Convergence of Exponentially Averaged Work Values." *Physical Review E* 73, no. 4 (April 5, 2006): 046105. https://doi.org/10.1103/PhysRevE.73.046105.
8. Yunger Halpern, Nicole, and Christopher Jarzynski. "Number of Trials Required to Estimate a Free-Energy Difference, Using Fluctuation Relations." *Physical Review E* 93, no. 5 (May 26, 2016): 052144. https://doi.org/10.1103/PhysRevE.93.052144.

CHAPTER 11 —·— RESOURCE THEORIES

1. Yirka, Bob. "Researchers Demonstrate Teleportation Using On-Demand Photons from Quantum Dots." Science X website. Accessed April 5, 2021. https://phys.org/news/2018-12-teleportation-on-demand-photons-quantum-dots.html.

2. Horodecki, Ryszard, Paweł Horodecki, Michał Horodecki, and Karol Horodecki. "Quantum Entanglement." *Reviews of Modern Physics* 81, no. 2 (June 17, 2009): 865–942. https://doi.org/10.1103/RevModPhys.81.865.
3. Chitambar, Eric, and Gilad Gour. "Quantum Resource Theories." *Reviews of Modern Physics* 91, no. 2 (April 4, 2019): 025001. https://doi.org/10.1103/RevModPhys.91.025001.
4. Marshall, Albert W., Ingram Olkin, and Barry C. Arnold. *Inequalities: Theory of Majorization and Its Applications*. Springer Series in Statistics. New York: Springer, 2011.
5. Ruch, Ernst, Rudolf Schranner, and Thomas H. Seligman. "The Mixing Distance." *Journal of Chemical Physics* 69, no. 1 (July 1, 1978): 386–92. https://doi.org/10.1063/1.436364.
6. Janzing, D., P. Wocjan, R. Zeier, R. Geiss, and T. Beth. "Thermodynamic Cost of Reliability and Low Temperatures: Tightening Landauer's Principle and the Second Law." *International Journal of Theoretical Physics* 39, no. 12 (December 1, 2000): 2717–53. https://doi.org/10.1023/A:1026422630734.
7. Horodecki, Michał, and Jonathan Oppenheim. "Fundamental Limitations for Quantum and Nanoscale Thermodynamics." *Nature Communications* 4, no. 1 (June 26, 2013): 2059. https://doi.org/10.1038/ncomms3059.
8. Gour, Gilad, David Jennings, Francesco Buscemi, Runyao Duan, and Iman Marvian. "Quantum Majorization and a Complete Set of Entropic Conditions for Quantum Thermodynamics." *Nature Communications* 9, no. 1 (December 17, 2018): 5352. https://doi.org/10.1038/s41467-018-06261-7.
9. Gour, Gilad, Markus P. Müller, Varun Narasimhachar, Robert W. Spekkens, and Nicole Yunger Halpern. "The Resource Theory of Informational Nonequilibrium in Thermodynamics." *Physics Reports* 583 (July 2, 2015): 1–58. https://doi.org/10.1016/j.physrep.2015.04.003.
10. Brandão, Fernando, Michał Horodecki, Nelly Ng, Jonathan Oppenheim, and Stephanie Wehner. "The Second Laws of Quantum Thermodynamics." *Proceedings of the National Academy of Sciences* 112, no. 11 (March 17, 2015): 3275–79. https://doi.org/10.1073/pnas.1411728112.
11. Yunger Halpern, Nicole, and Joseph M. Renes. "Beyond Heat Baths: Generalized Resource Theories for Small-Scale Thermodynamics." *Physical Review E* 93, no. 2 (February 18, 2016): 022126. https://doi.org/10.1103/PhysRevE.93.022126.
12. Yunger Halpern, Nicole. "Beyond Heat Baths II: Framework for Generalized Thermodynamic Resource Theories." *Journal of Physics A: Mathematical and Theoretical* 51, no. 9 (February 1, 2018): 094001.
13. Vaccaro, Joan A., and Stephen M. Barnett. "Information Erasure without

an Energy Cost." *Proceedings of the Royal Society A: Mathematical, Physical and Engineering Sciences* 467, no. 2130 (June 8, 2011): 1770–78. https://doi.org/10.1098/rspa.2010.0577.

14. Yunger Halpern, Nicole. "Toward Physical Realizations of Thermodynamic Resource Theories." In *Information and Interaction: Eddington, Wheeler, and the Limits of Knowledge*, edited by Ian T. Durham and Dean Rickles, 135–66. The Frontiers Collection. Cham, Germany: Springer International, 2017. https://doi.org/10.1007/978-3-319-43760-6_8.
15. Yunger Halpern, Nicole, Andrew J. P. Garner, Oscar C. O. Dahlsten, and Vlatko Vedral. "Introducing One-Shot Work into Fluctuation Relations." *New Journal of Physics* 17, no. 9 (September 11, 2015): 095003. https://doi.org/10.1088/1367-2630/17/9/095003.
16. Alhambra, Álvaro M., Lluis Masanes, Jonathan Oppenheim, and Christopher Perry. "Fluctuating Work: From Quantum Thermodynamical Identities to a Second Law Equality." *Physical Review* X 6, no. 4 (October 24, 2016): 041017. https://doi.org/10.1103/PhysRevX.6.041017.
17. Kucharski, Timothy J., Nicola Ferralis, Alexie M. Kolpak, Jennie O. Zheng, Daniel G. Nocera, and Jeffrey C. Grossman. "Templated Assembly of Photoswitches Significantly Increases the Energy-Storage Capacity of Solar Thermal Fuels." *Nature Chemistry* 6, no. 5 (May 2014): 441–47. https://doi.org/10.1038/nchem.1918.
18. Yunger Halpern, Nicole, and David T. Limmer. "Fundamental Limitations on Photoisomerization from Thermodynamic Resource Theories." *Physical Review A* 101, no. 4 (April 17, 2020): 042116. https://doi.org/10.1103/PhysRevA.101.042116.

CHAPTER 12 —·— THE UNSEEN KINGDOM

1. Yunger Halpern, Nicole, and Joseph M. Renes. "Beyond Heat Baths: Generalized Resource Theories for Small-Scale Thermodynamics." *Physical Review E* 93, no. 2 (February 18, 2016): 022126. https://doi.org/10.1103/PhysRevE.93.022126.
2. Yunger Halpern, Nicole. "Beyond Heat Baths II: Framework for Generalized Thermodynamic Resource Theories." *Journal of Physics A: Mathematical and Theoretical* 51, no. 9 (February 1, 2018): 094001.
3. Lostaglio, Matteo. "The Resource Theory of Quantum Thermodynamics." Master's thesis, Imperial College London, 2014.
4. Jaynes, E. T. "Information Theory and Statistical Mechanics." *Physical Review* 106, no. 4 (May 15, 1957): 620–30. https://doi.org/10.1103/PhysRev.106.620.

5. Jaynes, E. T. "Information Theory and Statistical Mechanics. II." *Physical Review* 108, no. 2 (October 15, 1957): 171–90. https://doi.org/10.1103/PhysRev.108.171.
6. Balian, Roger, and N. L. Balazs. "Equiprobability, Inference, and Entropy in Quantum Theory." *Annals of Physics* 179, no. 1 (October 1, 1987): 97–144. https://doi.org/10.1016/S0003-4916(87)80006-4.
7. Balian, Roger, Yoram Alhassid, and Hugo Reinhardt. "Dissipation in Many-Body Systems: A Geometric Approach Based on Information Theory." *Physics Reports* 131, no. 1 (January 1, 1986): 1–146. https://doi.org/10.1016/0370-1573(86)90005-0.
8. Lostaglio, Matteo, David Jennings, and Terry Rudolph. "Thermodynamic Resource Theories, Non-Commutativity and Maximum Entropy Principles." *New Journal of Physics* 19, no. 4 (April 6, 2017): 043008. https://doi.org/10.1088/1367-2630/aa617f.
9. Guryanova, Yelena, Sandu Popescu, Anthony J. Short, Ralph Silva, and Paul Skrzypczyk. "Thermodynamics of Quantum Systems with Multiple Conserved Quantities." *Nature Communications* 7, no. 1 (July 7, 2016): 12049. https://doi.org/10.1038/ncomms12049.
10. Yunger Halpern, Nicole, Philippe Faist, Jonathan Oppenheim, and Andreas Winter. "Microcanonical and Resource-Theoretic Derivations of the Thermal State of a Quantum System with Noncommuting Charges." *Nature Communications* 7, no. 1 (July 7, 2016): 12051. https://doi.org/10.1038/ncomms12051.
11. Yunger Halpern, Nicole, Michael E. Beverland, and Amir Kalev. "Noncommuting Conserved Charges in Quantum Many-Body Thermalization." *Physical Review E* 101, no. 4 (April 15, 2020): 042117. https://doi.org/10.1103/PhysRevE.101.042117.
12. Manzano, Gonzalo, Juan M. R. Parrondo, and Gabriel T. Landi. "Non-Abelian Quantum Transport and Thermosqueezing Effects." *ArXiv:2011.04560 [Cond-Mat, Physics:Quant-Ph]*, November 9, 2020. http://arxiv.org/abs/2011.04560.

CHAPTER 13 —·— ALL OVER THE MAP

1. Sørensen, Ole W. "A Universal Bound on Spin Dynamics." *Journal of Magnetic Resonance (1969)* 86, no. 2 (February 1, 1990): 435–40. https://doi.org/10.1016/0022-2364(90)90278-H.
2. Schulman, Leonard J., and Umesh V. Vazirani. "Molecular Scale Heat Engines and Scalable Quantum Computation." In *Proceedings of the Thirty-First Annual ACM Symposium on Theory of Computing (STOC)*, 322–29.

Atlanta, Georgia: Association for Computing Machinery, 1999. https://doi.org/10.1145/301250.301332.

3. Park, Daniel K., Nayeli A. Rodriguez-Briones, Guanru Feng, Robabeh Rahimi, Jonathan Baugh, and Raymond Laflamme. "Heat Bath Algorithmic Cooling with Spins: Review and Prospects." In *Electron Spin Resonance (ESR) Based Quantum Computing*, edited by Takeji Takui, Lawrence Berliner, and Graeme Hanson, 227–55. Biological Magnetic Resonance series, vol. 31. New York: Springer, 2016. https://doi.org/10.1007/978-1-4939-3658-8_8.
4. Boykin, P. Oscar, Tal Mor, Vwani Roychowdhury, Farrokh Vatan, and Rutger Vrijen. "Algorithmic Cooling and Scalable NMR Quantum Computers." *Proceedings of the National Academy of Sciences* 99, no. 6 (March 19, 2002): 3388–93. https://doi.org/10.1073/pnas.241641898.
5. Horowitz, Jordan M., and Todd R. Gingrich. "Thermodynamic Uncertainty Relations Constrain Non-Equilibrium Fluctuations." *Nature Physics* 16, no. 1 (January 2020): 15–20. https://doi.org/10.1038/s41567-019-0702-6.
6. Pietzonka, Patrick, Andre C Barato, and Udo Seifert. "Universal Bound on the Efficiency of Molecular Motors." *Journal of Statistical Mechanics: Theory and Experiment* 2016, no. 12 (December 30, 2016): 124004. https://doi.org/10.1088/1742-5468/2016/12/124004.
7. Ptaszyński, Krzysztof. "Coherence-Enhanced Constancy of a Quantum Thermoelectric Generator." *Physical Review B* 98, no. 8 (August 20, 2018): 085425. https://doi.org/10.1103/PhysRevB.98.085425.
8. Agarwalla, Bijay Kumar, and Dvira Segal. "Assessing the Validity of the Thermodynamic Uncertainty Relation in Quantum Systems." *Physical Review B* 98, no. 15 (October 26, 2018): 155438. https://doi.org/10.1103/PhysRevB.98.155438.
9. Macieszczak, Katarzyna, Kay Brandner, and Juan P. Garrahan. "Unified Thermodynamic Uncertainty Relations in Linear Response." *Physical Review Letters* 121, no. 13 (September 24, 2018): 130601. https://doi.org/10.1103/PhysRevLett.121.130601.
10. Guéry-Odelin, D., A. Ruschhaupt, A. Kiely, E. Torrontegui, S. Martínez-Garaot, and J.G. Muga. "Shortcuts to Adiabaticity: Concepts, Methods, and Applications." *Reviews of Modern Physics* 91, no. 4 (October 24, 2019): 045001. https://doi.org/10.1103/RevModPhys.91.045001.
11. Albash, Tameem, and Daniel A. Lidar. "Adiabatic Quantum Computation." *Reviews of Modern Physics* 90, no. 1 (January 29, 2018): 015002. https://doi.org/10.1103/RevModPhys.90.015002.
12. Bäumer, Elisa, Matteo Lostaglio, Martí Perarnau-Llobet, and Rui Sampaio.

"Fluctuating Work in Coherent Quantum Systems: Proposals and Limitations." In *Thermodynamics in the Quantum Regime: Fundamental Aspects and New Directions*, edited by Felix Binder, Luis A. Correa, Christian Gogolin, Janet Anders, and Gerardo Adesso, 275–300. Fundamental Theories of Physics series. Cham, Germany: Springer International, 2018. https://doi.org/10.1007/978-3-319-99046-0_11.

13. Wakakuwa, Eyuri. "Operational Resource Theory of Non-Markovianity." *ArXiv:1709.07248 [Quant-Ph]*, October 3, 2017. http://arxiv.org/abs/1709.07248.
14. Pezzutto, Marco, Mauro Paternostro, and Yasser Omar. "Implications of Non-Markovian Quantum Dynamics for the Landauer Bound." *New Journal of Physics* 18, no. 12 (December 15, 2016): 123018. https://doi.org/10.1088/1367-2630/18/12/123018.
15. Mehboudi, Mohammad, Anna Sanpera, and Luis A Correa. "Thermometry in the Quantum Regime: Recent Theoretical Progress." *Journal of Physics A: Mathematical and Theoretical* 52, no. 30 (July 26, 2019): 303001. https://doi.org/10.1088/1751-8121/ab2828.
16. Jevtic, Sania, David Newman, Terry Rudolph, and T. M. Stace. "Single-Qubit Thermometry." *Physical Review A* 91, no. 1 (January 22, 2015): 012331. https://doi.org/10.1103/PhysRevA.91.012331.
17. Stace, Thomas M. "Quantum Limits of Thermometry." *Physical Review A* 82, no. 1 (July 30, 2010): 011611. https://doi.org/10.1103/PhysRevA.82.011611.

CHAPTER 14 —·— STEPPING OFF THE MAP

1. "Black Holes." Science Mission Directorate, NASA Science website. Accessed April 5, 2021. https://science.nasa.gov/astrophysics/focus-areas/black-holes.
2. Hawking, S. W. "Particle Creation by Black Holes." *Communications in Mathematical Physics* 43, no. 3 (August 1, 1975): 199–220. https://doi.org/10.1007/BF02345020.
3. Kitaev, Alexei. "A Simple Model of Quantum Holography (Part 1)." Conference presentation at Entanglement in Strongly Correlated Quantum Matter, Kavli Institute for Theoretical Physics, April 7, 2015. https://online.kitp.ucsb.edu/online/entangled15/kitaev/.
4. Larkin, A. I., and Yu. N. Ovchinnikov. "Quasiclassical Method in the Theory of Superconductivity." *Soviet Journal of Experimental and Theoretical Physics* 28 (June 1, 1969): 1200.
5. "This Month in Physics History: Circa January 1961: Lorenz and the Butterfly Effect." *APS News* 12, no. 1 (January 2003). http://www.aps.org/publications/apsnews/200301/history.cfm.

6. Yunger Halpern, Nicole. "Jarzynski-like Equality for the Out-of-Time-Ordered Correlator." *Physical Review A* 95, no. 1 (January 17, 2017): 012120. https://doi.org/10.1103/PhysRevA.95.012120.
7. Solinas, P., and S. Gasparinetti. "Full Distribution of Work Done on a Quantum System for Arbitrary Initial States." *Physical Review E* 92, no. 4 (October 23, 2015): 042150. https://doi.org/10.1103/PhysRevE.92.042150.
8. Campisi, Michele, and John Goold. "Thermodynamics of Quantum Information Scrambling." *Physical Review E* 95, no. 6 (June 20, 2017): 062127. https://doi.org/10.1103/PhysRevE.95.062127.
9. Touil, Akram, and Sebastian Deffner. "Quantum Scrambling and the Growth of Mutual Information." *Quantum Science and Technology* 5, no. 3 (May 26, 2020): 035005. https://doi.org/10.1088/2058-9565/ab8ebb.
10. Arute, Frank, Kunal Arya, Ryan Babbush, Dave Bacon, Joseph C. Bardin, Rami Barends et al. "Quantum Supremacy Using a Programmable Superconducting Processor." *Nature* 574, no. 7779 (October 2019): 505–10. https://doi.org/10.1038/s41586-019-1666-5.

EPILOGUE —·— WHERE TO NEXT?

1. Choi, Joonhee, Hengyun Zhou, Renate Landig, Hai-Yin Wu, Xiaofei Yu, Stephen E. Von Stetina et al. "Probing and Manipulating Embryogenesis via Nanoscale Thermometry and Temperature Control." *Proceedings of the National Academy of Sciences* 117, no. 26 (June 30, 2020): 14636–41. https://doi.org/10.1073/pnas.1922730117.
2. Fujiwara, Masazumi, Simo Sun, Alexander Dohms, Yushi Nishimura, Ken Suto, Yuka Takezawa et al. "Real-Time Nanodiamond Thermometry Probing in Vivo Thermogenic Responses." *Science Advances* 6, no. 37 (September 1, 2020): eaba9636. https://doi.org/10.1126/sciadv.aba9636.

INDEX

Page numbers followed by *n* indicate footnotes.